Monographs in Theoretical Computer Science

An EATCS Series

Monographs in Theoretical Computer Science
An EATCS Series

Ariel Gabizon

Deterministic Extraction from Weak Random Sources

Dr. Ariel Gabizon
University of Texas at Austin
Dept. Computer Science
1 University Station C0500
Taylor Hall
78712-1188 Austin, TX Texas
USA
ariel.gabizon@gmail.com

Series Editors
Prof. Dr. Wilfried Brauer
Institut für Informatik der TUM
Boltzmannstr. 3
85748 Garching, Germany
brauer@informatik.tu-muenchen.de

Prof. Dr. Grzegorz Rozenberg
Leiden Institute of Advanced
Computer Science
University of Leiden
Niels Bohrweg 1
2333 CA Leiden, The Netherlands
rozenber@liacs.nl

Prof. Dr. Juraj Hromkovič
ETH Zentrum
Department of Computer Science
Swiss Federal Institute of Technology
8092 Zürich, Switzerland
juraj.hromkovic@inf.ethz.ch

Prof. Dr. Arto Salomaa
Turku Centre of Computer Science
Lemminkäisenkatu 14 A
20520 Turku, Finland
asalomaa@utu.fi

ISSN 1431-2654
ISBN 978-3-642-26538-9 ISBN 978-3-642-14903-0 (eBook)
DOI 10.1007/978-3-642-14903-0
Springer Heidelberg Dordrecht London New York

ACM Computing Classification (1998): F.1, F.2, G.2, G.3

Cover design: KuenkelLopka GmbH, Heidelberg

Printed on acid-free paper

Springer is part of Springer Science+Business Media (www.springer.com)

Preface

Roughly speaking, a *deterministic extractor* is a function that 'extracts' almost perfect random bits from a 'weak random source' - a distribution that contains some entropy but is far from being truly random. In this book we explicitly construct deterministic extractors and related objects for various types of sources. A basic theme in this book is a methodology of recycling randomness that enables increasing the output length of deterministic extractors to near-optimal length. Our results are as follows.

Deterministic Extractors for Bit-Fixing Sources An (n, k)-*bit-fixing source* is a distribution X over $\{0,1\}^n$ such that there is a subset of k variables in $X_1, \ldots, X_n$ that are uniformly distributed and independent of each other, and the remaining $n-k$ variables are fixed in advance to some (unknown) constants. We give constructions of deterministic bit-fixing source extractors that extract $(1-o(1))k$ bits whenever $k > (\log n)^c$ for some universal constant $c > 0$. Thus, our constructions extract almost all the randomness from bit-fixing sources and work even when k is small. Our technique gives a general method to transform deterministic bit-fixing source extractors that extract few bits into extractors which extract almost all the bits.

Deterministic Extractors for Affine Sources over Large Fields An (n, k)-*affine source* over a finite field $\mathbb{F}$ is a random variable $X = (X_1, ..., X_n) \in \mathbb{F}^n$, that is uniformly distributed over an (unknown) k-dimensional affine subspace of $\mathbb{F}^n$. There has been much interest lately in extractors for affine sources over $\mathbb{F}_2$. It can be shown that a random function $D : \{0,1\}^n \mapsto \{0,1\}$ is with high probability an extractor for (n, k)-affine sources over $\mathbb{F}_2$ whenever $k \geq 3 \cdot \log n$. The best explicit construction due to Bourgain [10] works when $k = \delta \cdot n$ for constant δ.

We focus on the case of a *large* field, specifically, a field of size n^c for constant $c > 0$, i.e., a field size that is polynomially large in the dimension of the space. When working with a field of size larger than n^{20} we show how to deterministically extract practically all the randomness from an (n, k)-affine source for any $k \geq 2$.

Extractors and Rank Extractors for Polynomial Sources We construct explicit deterministic extractors from *polynomial sources*, namely from distributions sampled by low degree multivariate polynomials over finite fields. This naturally generalizes previous work on extraction from affine sources (which are degree 1 polynomials).

The first step in our construction is a construction of *rank extractors*, which are polynomial mappings that "extract" the algebraic rank from any system of low-degree polynomials. More precisely, for any n polynomials, k of which are algebraically independent, a rank extractor outputs k algebraically independent polynomials of slightly higher degree.

We then use theorems of Wooley and Bombieri from algebraic geometry, which enable us to extract a constant fraction of the randomness from 'full rank' polynomial sources when the field is exponentially large in the degrees of the defining polynomials.

Increasing the Output Length of Zero-Error Dispersers A zero-error disperser for a family of weak random sources is a function that guarantees the output distribution will have full support for any source in the family. We develop a general method of increasing the output length of zero-error dispersers. We use this method to significantly improve previous constructions. More specifically, we obtain zero-error dispersers for 2-independent sources, bit-fixing sources and affine sources over large fields with output length $\Omega(k)$ where k is the min-entropy of the source.

April 2010 Ariel Gabizon

Acknowledgements

This monograph is based on my doctoral dissertation, written under the supervision of Ran Raz and Ronen Shaltiel, and submitted to the Weizmann Institute in Israel in June 2008. Above all, I'd like to thank my advisors Ran and Ronen for showing a lot of faith in me and my abilities when I started out.

Ran showed a very positive attitude from the start and gave the feeling that 'anything is possible' and 'everything is going great'. This created a very comfortable and carefree environment for doing research for me, where I felt what I had already done was great and I was free to try out anything in the future. This feeling was preserved due to Ran's ability to remain open-minded and patient throughout the years. For me, Ran is an incredible model for clarity of thought, effective thought and patience with people (and these qualities indeed manifested in a few crucial points of the research).

This thesis began with an idea of Ronen Shaltiel on the possibility of 'recycling randomness' in a certain situation. This idea became a central theme to which almost all results in this thesis are connected. Thus, I owe this thesis to Ronen in a very concrete sense. I am grateful to Ronen for the great willingness, enthusiasm and modesty with which he shares his knowledge. Indeed, it was at a colloquium lecture of Ronen at the Hebrew University that I first realized how cool pseudorandomness and complexity theory are. Ronen has helped direct and focus me with his 'stock' of great research directions, and his almost magical ability to take any vague idea you bring, develop it, find where it may be useful, all the while making it seem like it is still just *your* idea!

During the last year and a half of the thesis I spent a lot of time learning about function fields and algebraic geometric codes. This became much more effective in the last few months thanks to Irit Dinur who took active interest in the learning process, invested a lot of time, and created an environment where I was forced, to a certain extent, to really know what I was talking about.

I'd like to thank Chris Umans for a great two weeks spent working with him at Caltech. He really taught me what it is like to 'brainstorm' with someone and inspired me with his evident passion for research. Thanks to my coauthors Zeev Dvir and Avi Wigderson who participated in the research presented here (Chapter 4 is joint research with them). I had a fun and fruitful time in China thanks to Andy Yao's inviting me to China Theory Week 2007. Thanks to my PhD committee Oded Goldreich and Omer Reingold. Thanks again to Oded Goldreich for useful comments on all chapters in this monograph.

There was a talented theory group at Weizmann thanks to Zeev Dvir, Dana Moshkovitz, Amir Yehudayoff, Anup Rao (who visited for a summer), Adi Shraibman, Danny Harnik, Tal Moran, Ronen Gradwohl, Gil Segev, Iftach Haitner, Noam Livne, Yuval Emek, Asaf Nussbaum, Shachar Lovett, Erez Kantor, Gillat Kol, Or Meir, Zvika Brakerski, Omer Kadmiel, Ran Halprin, Chandan Dubey and Guy Kindler. I feel grateful that I was able to spend my time in a field full of creativity, depth and beauty. I think that doing this for a living is really a privilege.

The Weizmann Institute will always have a special place in my heart. With every day that passed, through meeting impressive and inspiring people, and enjoying the great freedom we have to develop both personally and professionally, I became more convinced what a great place Weizmann is. I could also say Weizmann is 'in my blood', as I was born to two Weizmann Ph.D. students who met at the Wolfson building right across from the building where I worked all those years.

My period at Weizmann started rather spontaneously. I actually started my graduate studies elsewhere. Towards the end of the first semester, while browsing the Weizmann web site and seeing the courses for the next semester, I had a strong feeling I should move to Weizmann immediately, and study complexity theory.

A possible definition of 'fate' might be 'the things that start working out when you give a small nudge in their direction'. A friend of mine, who was already a Weizmann student said I could move into a spare room in his apartment in Rehovot. I am very grateful to him, as knowing my laziness in these matters and my fear of unknown roommates, and with only a few weeks left till the beginning of the second semester, without his offer there is a good chance I would have forsaken the plan. I also want to thank Uri Feige who was my 'first contact' with the department. Uri was then in charge of teaching, and I phoned him to ask about the possibility of taking courses as an informal student. He was extremely kind and patient with my questions and helped me relax and feel that 'it was going to be OK'.

I distinctly remember the day I moved to Rehovot, arriving that evening with my luggage at my friend's apartment. For me, that moment was a sharp turning point, setting in motion a new course in my life.

To my family and all the good people I met during those 5 years, thank you for your support. May you be happy.

Contents

Chapter 1

Introduction

1.1 Organization of This Book

This monograph is concerned with the explicit construction of deterministic extractors and related objects. The current chapter gives an introduction to extractors and a sketch of the techniques and results of this work. We begin in Section 1.2 with what may be called the 'classical motivation' for extractors. Section 1.3 describes other motivations for deterministic extractors. In Section 1.4 we describe some of the main techniques used in this book. The subsequent chapters describe our results in detail. Each chapter is self-contained, perhaps with the exception of some basic background given in this introductory chapter. For a concise statement of all results, see the summary in the preface.

1.2 The Classical Story

Randomness is an important resource. For example, many algorithms and cryptographic applications require random bits in order to be executed. Outside of computer science, other scientists, from archaeologists to biologists to physicists, need random numbers to simulate physical processes and survey large groups in experiments. A poet has recognized the power of randomness in the following verse:

Oh, many a shaft at random sent
Finds mark the archer little meant!
And many a word at random spoken
May soothe, or wound, a heart that's broken! - Sir Walter Scott

This raises the question of *how to obtain random bits.* In the beginning of the last century, scientists who needed random numbers actually tossed

A. Gabizon, *Deterministic Extraction from Weak Random Sources*,
Monographs in Theoretical Computer Science. An EATCS Series,
DOI 10.1007/978-3-642-14903-0_1,

coins or threw cubes. Here are a few methods used today by Web sites that offer free random numbers:

- In www.fourmilab.ch/hotbits, the decay time of radioactive particles is used (visit the site for a very readable and fascinating explanation of the physics behind how it works).
- In www.random.org, a radio is tuned to a frequency at which no one is broadcasting and the atmospheric noise is recorded and then processed to remove some correlations.
- In www.stat.fsu.edu/~geo/diehard.html, Dr. George Marsaglia has combined white noise from classical music and rap CDs (together with some other stuff).

A question that arises is: How can we be sure that these methods produce *truly random bits*? By truly random bits we mean that each bit is 1 with probability 1/2 and is independent of all the other bits in the sequence generated. The first method, assuming the validity of quantum mechanics, is theoretically guaranteed to produce truly random bits. The other two methods described, and most methods used, offer no such guarantee. The quality of the numbers produced by these methods is checked empirically by statistical tests but there is no reason to assume that these methods produce truly random bits. To summarize, it seems that while it is possible to get truly random bits from nature, we have an abundance of very cheap and accessible "weak random sources" and it would be very convenient to be able to use these sources.
Suppose, then, that we only want to make a weak assumption about the quality of randomness in our source. Specifically, we assume that it belongs to some class $\mathcal{C}$ of "weak random sources". A *deterministic extractor* is a function that enables us to "extract" (almost) truly random bits from such a "weak random source". Formally

Definition 1.1 (deterministic extractor). *Let $\mathcal{C}$ be a class of distributions on $\{0,1\}^n$. A function $E : \{0,1\}^n \to \{0,1\}^m$ is a deterministic ϵ-extractor for $\mathcal{C}$ if for every distribution $X \in \mathcal{C}$ the distribution $E(X)$ (obtained by sampling x from X and computing $E(x)$) is ϵ-close to the uniform distribution on m bit strings.*[1]

What Class $\mathcal{C}$ of "weak random sources" should we consider? The largest class we can hope to extract randomness from is the class of *high min-entropy* sources originally defined in [14]. A high min-entropy source is guaranteed not to output any particular string with large probability. No further assumption is made about the structure of the source.

[1]Two distributions P and Q over $\{0,1\}^m$ are ϵ-close (denoted by $P \stackrel{\epsilon}{\sim} Q$) if for every event $A \subseteq \{0,1\}^m$, $|P(A) - Q(A)| \leq \epsilon$.

Definition 1.2. *The min-entropy of a distribution X over $\{0,1\}^n$ is*

$$H_\infty(X) = min_{x \in \{0,1\}^n} \log(1/\Pr(x)).$$

X is called a k-source if it has min-entropy k.

1.2.1 Seeded Extractors

We would like to have a deterministic extractor for the class of all k-sources. Unfortunately, this isn't possible even when $k = n - 1$. This leads to the notion of a *seeded extractor* (see the surveys [62, 46]). A seeded extractor (sometimes simply called extractor) uses a short random 'seed' that 'helps it' extract the randomness out of the source. In contrast to a deterministic extractor, a seeded extractor can extract randomness from any k-source.

Definition 1.3 (seeded extractors for high min-entropy sources). *A function $E : \{0,1\}^n \times \{0,1\}^d \to \{0,1\}^m$ is a (k, ϵ)-extractor if for any k-source X the distribution $E(X, U_d)$ is ϵ-close to U_m.*

It should be noted that in addition to its original motivation for extracting randomness, the notion of a seeded extractor has turned out to be a very natural and useful one. For example, seeded extractors have been used in constructions of expander graphs and pseudorandom generators and have been shown to be equivalent in some sense to certain types of error correcting codes and samplers (again, see the surveys [62, 46] for more details).

1.2.2 Deterministic Extraction for Restricted Classes

As mentioned above, there are no deterministic extractors for high min-entropy sources. However, we can construct deterministic extractors for more restricted classes of distributions. Here are some examples of such classes.

- A sequence of *independent* tosses of a *biased* coin [71].
- Bit-fixing sources [15, 16, 37].
- Sources consisting of few independent blocks [14, 19, 3, 4, 53, 50, 9, 5].
- Sources samplable by small circuits [67].

1.3 Other Motivations

We now describe motivations for deterministic extractors in the context of theoretical computer science.

Deterministic extractors in derandomization and pseudorandomness The field of *derandomization* deals with taking computational tasks that currently require the use of randomness, and trying to achieve them using less randomness, or ideally, none at all. A useful tool in derandomization is constructing *pseudorandom objects*, i.e., objects that have properties that a random object would have with high probability. Deterministically constructing a pseudorandom object is also a derandomization problem in itself — it derandomizes the procedure that simply chooses a random object. Maybe the most basic motivation for extractors comes from this setting. Given any 'not too large' family of weak random sources, it is easy to see that a random function is with high probability a deterministic extractor for this family — simply take a union bound over all events of the form 'not being a good extractor for X' for a distribution X in the family.

Thus, deterministic extractors can be thought of as pseudorandom objects and explicitly constructing them can be thought of as a derandomization problem.

We note that proving lower bounds for functions — perhaps the central goal of complexity theory — can be viewed as a problem of explicitly constructing pseudorandom objects: For example, a random function with high probability does not have polynomial size circuits. Finding a function in $\mathcal{NP}$ with this property would show $\mathcal{NP} \not\subseteq \mathcal{P}/poly$, and in particular $\mathcal{P} \neq \mathcal{NP}$. Thus, it may be hoped that understanding better how to construct functions with 'simple' pseudorandom properties — e.g., the property of being an extractor for a certain family of sources, will eventually help in understanding how to construct functions with the 'ultimate' pseudorandom property, i.e., having high circuit complexity.

Useful pseudorandom properties of deterministic extractors Let's try to see what specific properties deterministic extractors have that are useful and natural. Assume our class $\mathcal{C}$ of weak random sources consists of uniform distributions over subsets X of $\{0,1\}^n$. Suppose E is a deterministic extractor for $\mathcal{C}$ that extracts one bit. E is a coloring of $\{0,1\}^n$ (with two colors) such that every subset X is colored in a balanced way. In particular, no subset X is monochromatic. For example, our extractors for affine sources, described in Chapter 3, give explicit colorings of vector spaces such that each line is colored in a balanced way. From another point of view, a deterministic extractor can be seen as a function that gets a (completely) random input and produces an output that looks random even to an adversary who has learned something about the input. For example, our extractors for bit-fixing sources[26], described in Chapter 2, are functions that output almost k bits that look random to an adversary who knows $n-k$ bits out of the n-bit input. Such functions are useful in "exposure resilient cryptography" [57, 11, 12, 20, 18][2].

---

[2] The parameters of our construction are not good enough for some of the applications in this book.

The power of weak randomness A central issue in complexity theory is discovering what effect the availability of randomness has on the computational power. One specific question is what happens to the computational power when we have access to a weak random source instead of truly random bits. Explicit constructions of seeded extractors with logarithmic seed length show us that probabilistic algorithms can be run using any high min-entropy source by iterating over all seeds and taking the majority answer. However, this approach obviously won't work in cryptographic settings. In fact, a result of Dodis et al. [74] shows that many cryptographic protocols cannot be executed using an SV source (SV sources [58] are a sub-class of high min-entropy sources). Constructing an explicit deterministic extractor (that extracts enough bits) for a certain weak random source shows us that access to this type of weak randomness gives as much computational power as truly random bits (in any scenario where the parties are powerful enough to compute the extractor).

1.4 Techniques — the Recycling Paradigm

A central technique in this thesis is a general methodology of 'recycling randomness'. More specifically, we show that in certain instances we can use randomness deterministically extracted from a source as a seed for a seeded function that is applied *on the same source*, and get almost the same output distribution we would have gotten by applying the seeded function with a seed that is *independent* of the source. In this section we explain the recycling paradigm and on the way sketch the results we obtain using it.

1.4.1 A Simple Example

We give a simple example of the recycling paradigm using the following toy problem. Suppose you are given two independent bit-strings X_A and X_B each of length k. You are guaranteed that one of the strings is completely random, and that the other string contains one random bit and all other bits in that string are fixed to some constants (some may be fixed to 0 and some to 1). You have no additional randomness of your own. Your goal is to choose the random string with probability (at least) one half without 'ruining it'. That is, the random string has to stay random when conditioning on the event that you chose it.

Solution Compute Z = the XOR (i.e. sum mod 2) of all bits in X_A and X_B. If $Z = 1$ return X_A. If $Z = 0$ return X_B.

Proof. Note that Z is obviously a random bit, as it is the XOR of random bits and constant bits. So, with probability 1/2 we choose the random string. Assume w.l.o.g. that X_A is the random string. The question is whether X_A is still random *conditioned on the event that we chose it*, i.e., conditioned on

$Z = 1$. Let's fix a possible value $b \in \{0,1\}^k$ for X_A. The first thing to notice is that conditioning on $X_A = b$ does not affect Z. That is,

$$\Pr(Z = 1 | X_A = b) = \frac{1}{2} = \Pr(Z = 1).$$

This is because after this conditioning Z is still the XOR of random bits (the random bit from X_B) and constant bits, and thus still random. Using the Bayes formula for conditional probabilities we can 'reverse the conditoning' and get that conditioning on Z does not affect X_A:

$$\Pr(X_A = b | Z = 1) = \frac{\Pr(X_A = b \wedge Z = 1)}{\Pr(Z = 1)}$$

$$= \frac{\Pr(Z = 1 | X_A = b) \cdot \Pr(X_A = b)}{\Pr(Z = 1)} = \frac{\Pr(Z = 1) \cdot \Pr(X_A = b)}{\Pr(Z = 1)}$$

$$= \Pr(X_A = b).$$

Thus X_A is still uniformly distributed when conditioning on $Z = 1$. □

Let $X \triangleq X_A \circ X_B$. We saw that we were able to use a random bit that was a *function* of X as a seed for a function applied on X itself — namely, the function that chooses which half of the bits to output — and get the same output distribution we would have gotten when using a random seed that is *independent of* X.

1.4.2 The General Principle and the Application for Affine Sources

Let us put the above example in the context of bit-fixing sources.

Definition 1.4 (bit-fixing sources). *A distribution X over $\{0,1\}^n$ is an (n,k)-bit-fixing source if there exists a subset $S = \{i_1, \cdots, i_k\} \subseteq \{1, \ldots, n\}$ such that $X_{i_1}, X_{i_2}, \cdots, X_{i_k}$ is uniformly distributed over $\{0,1\}^k$ and for every $i \notin S$, X_i is constant.*

Note that we can think of $X = X_A \circ X_B$ as a bit-fixing source with $k+1$ random bits. Think of the function that chooses either X_A or X_B according to a one-bit seed as a function $E : \{0,1\}^{2k} \times \{0,1\} \mapsto \{0,1\}^k$. Denote the function that computes the parity of all bits by $D : \{0,1\}^{2k} \mapsto \{0,1\}$. In this notation, in the example above our solution was to compute $E(X, D(X))$. The property that we used was that for any $b \in \{0,1\}^k$, given $X_A = b$, which is equivalent to $E(X,1) = b$, X was still a bit-fixing source, and therefore $D(X | E(X,1) = b)$ was uniform. This allowed us to conclude that the distribution $E(X,1) | D(X) = 1$ was identical to $E(X,1)$.

In the next subsection we will see how in a similar spirit one can use the recycling paradigm for extracting randomness from general bit-fixing sources.

Using this logic we can deduce a general theorem for settings in which we can use correlated randomness.

Theorem 1.5 (composition theorem). *Let $\mathcal{C}$ be a class of weak random sources*[3] *on $\{0,1\}^n$ and let $X \in \mathcal{C}$. Let*

- *$D : \{0,1\}^n \mapsto \{0,1\}^d$ be a deterministic extractor for $\mathcal{C}$ with error $\epsilon = 0$.*
- *$E : \{0,1\}^n \times \{0,1\}^d \mapsto \{0,1\}^m$ be a function such that for every $a \in \{0,1\}^d, b \in \{0,1\}^m$*

$$(X|E(X,a) = b) \in \mathcal{C}.$$

Then,

$$E(X, D(X)) \sim E(X, U_d).$$

Remark 1.1. *We note that*

- *The actual theorem used works under more relaxed conditions: The error of D does not have to be 0, and in the condition on E, 'for every $a \in \{0,1\}^d$', can be replaced by "for most' $a \in \{0,1\}^d$'. In this introduction we intentionally give a simpler version to explain the idea.*
- *Again, to simplify the presentation, in the coming examples we make (unrealistic) simplifying assumptions to enable working with the condition 'for every $a \in \{0,1\}^d$' rather than "for most' $a \in \{0,1\}^d$'.*

Here is a nice application of this theorem used in [25]. Let $\mathbb{F}_q$ be some finite field and n be some integer. For the sake of this discussion, by an *affine source of dimension* k we mean a uniform distribution over an affine subspace $X \subseteq \mathbb{F}_q^n$ of dimension k. Let $\mathcal{C}$ be the class of affine sources of dimension at least 1. Let $D : \mathbb{F}_q^n \mapsto \{0,1\}^d$ be a deterministic extractor for $\mathcal{C}$ with error $\epsilon = 0$. Let $E : \mathbb{F}_q^n \times \{0,1\}^d \mapsto \mathbb{F}_q^{k-1}$ be a seeded extractor for affine sources of dimension k with error ϵ that is *linear* for any fixed seed. That is, for any $a \in \{0,1\}^d$, $T(x) \triangleq E(x,a)$ is a linear function of x. Then, we have

Theorem 1.6. *The function $F : \mathbb{F}_q^n \mapsto \mathbb{F}_q^{k-1}$ defined by*

$$F(x) \triangleq E(x, D(x))$$

is a deterministic extractor for affine sources of dimension k with error ϵ.

Proof. Let X be an affine source of dimension k. Fix any $a \in \{0,1\}^d$ and $b \in \mathbb{F}_q^{k-1}$. Then the condition $E(X,a) = b$ imposes $k-1$ affine conditions on

[3] The theorem's logic originated from [26]. However, the theorem for a general class $\mathcal{C}$ as presented here was explicitly written only by Shaltiel in [63].

the support of X. Hence, $(X|E(X,a) = b)$ is still an affine source and has dimension at least 1.[4] Using the Composition Theorem we have

$$F(X) = E(X, D(X)) \sim E(X, U_d)$$

and the later distribution is ϵ-close to uniform. □

In Chapter 3, we construct a deterministic extractor for affine sources that extracts a few bits, and a seeded extractor for affine sources, which is linear for any fixed seed, that extracts many bits. We can then use the above theorem to construct a deterministic extractor for affine sources that extracts many bits.

1.4.3 The Recycling Paradigm in Bit-Fixing Sources

The recycling paradigm was first used in [26] to increase the output length of deterministic extractors for bit-fixing sources. Very briefly, the idea is as follows: We use a sampler to choose a subset of the source bits that contains (with high probability) 'most but not all' of the random source bits. We then apply a seeded extractor on this subset of bits. Theorem 1.5 implies that we can use the sampler and extractor with a seed extracted from the source, and get the same distribution as when using a seed independent from the source. Here is a more detailed informal sketch: Let X be an (n, k)-bit-fixing source. What functions T have the property that $(X|T(X) = b)$ is still a bit-fixing source?

Let us refer to the indices $i \in [n]$ such that X_i is random as the *good indices of* X. Let $I \subseteq [n]$ be a subset of indices that contains $3k/4$ of the good indices of X. For $x \in \{0,1\}^n$ denote by x_I the restriction of x to the indices in I. Let T be the function $T(x) \triangleq x_I$. Then $(X|T(X) = b)$ is an $(n, k/4)$-bit-fixing source, as conditioning on $T(X) = b$ simply 'fixes' $3k/4$ random bits. Similarly, if $T(x)$ was some function of x_I then $(X|E(X,a) = b)$ is either an $(n, k/4)$-bit-fixing source or a convex combination of such sources, but for simplicity let's ignore the second option.

When constructing an extractor that should work for *all* (n, k)-bit-fixing sources we do not know in advance which indices of X are good. However, we can use a sampler S to choose a subset $I \subseteq [n]$ that with high probability contains about $3k/4$ good indices of X. We could then apply a seeded extractor E' on X_I. Let E be the function resulting from choosing I by S and then applying E' on X_I. Formally,

$$E(x, a = (a_1, a_2)) \triangleq E'(x_{S(a_1)}, a_2).$$

[4]For simplicity, we are ignoring the alternative case were $(X|E(X,a) = b)$ has empty support. In [25] we use a function E for which we show this does not happen for most $a \in \{0,1\}^d$.

For simplicity, let's assume that S always returns a subset I containing exactly $3k/4$ good indices of X. So, we have that for any $a \in \{0,1\}^d$ and $b \in \{0,1\}^m$, $(X|E(X,a) = b)$ is an $(n, k/4)$-bit-fixing source. Thus, if D is a deterministic extractor for $(n, k/4)$-bit-fixing sources it follows from the Composition Theorem that

$$E(X, D(X)) \sim E(X, U_d)$$

and the latter distribution is close to uniform. Thus, by defining $F(x) \triangleq E(x, D(x))$ we get a deterministic extractor with E's output length. See Chapter 2 for the full details.

1.4.4 The Recycling Paradigm for Zero-Error Dispersers

A *zero-error disperser* for a class $\mathcal{C}$ of distributions over $\{0,1\}^n$ is a function $D : \{0,1\}^n \mapsto \{0,1\}^m$ such that for any $X \in \mathcal{C}$

$$D(X) = \{0,1\}^m.$$

Zero-error dispersers are natural objects when viewed as colorings: Assume our class $\mathcal{C}$ consists of uniform distributions over subsets of $\{0,1\}^n$. A zero-error disperser $D : \{0,1\}^n \mapsto \{0,1\}^m$ for $\mathcal{C}$ is a coloring of $\{0,1\}^n$ such that any such subset contains all 2^m colors. A deterministic extractor $D : \{0,1\}^n \mapsto \{0,1\}^m$ is a zero-error disperser when it has error $\epsilon < 2^{-m}$. Alternatively, we can truncate the output length of D to $m = \log(1/\epsilon)$ to get a zero-error disperser. We would like to use the recycling method to get zero-error dispersers with longer output length. This does not work directly. One reason is the following. When using the Composition Theorem for extractors with a deterministic extractor $D : \{0,1\}^n \mapsto \{0,1\}^d$ with error ϵ_1 and a seeded extractor $E : \{0,1\}^n \times \{0,1\}^d \mapsto \{0,1\}^m$ with error ϵ_2, the new deterministic extractor $F(x) \triangleq E(x, D(x))$ will have error at least $\epsilon_1 + \epsilon_2$.[5] Thus, for F to be a zero-error disperser it is necessary that both $\epsilon_1 < 2^{-m}$ and $\epsilon_2 < 2^{-m}$. However, seeded extractors typically[6] have seed length $d \geq \log(1/\epsilon_2) \geq m$. which means D's output length needs to be at least m. As $\epsilon_1 < 2^{-m}$ D is already a zero-error disperser with output length m. Thus, we have not gained anything by this transformation.

In Chapter 5, based on [27], we use a different composition theorem that basically argues about 'sets rather than distributions' and enables showing the composed function will have full support *without showing it is close to uniform.*

[5] Note that we have not actually stated such a theorem in this exposition; as in Theorem 1.5, we assume for simplicity that D has error $\epsilon_1 = 0$.

[6] Seeded extractors for general sources *must* have seed length at least $\log(1/\epsilon)$. In this thesis we also use seeded extractors for restricted sources, e.g., affine sources and bit-fixing sources. For such sources obviously there is no such lower bound on seed length as we have *seedless* extractors in these cases. However, the examples we have come up with, and it seems any 'natural' seeded extractor for these sources, have seed length $\geq \log(1/\epsilon)$.

Here is a simplified version. For a distribution X we denote its support by $\mathrm{Supp}(X)$.

Theorem 1.7 (composition theorem for zero-error dispersers). *Let $\mathcal{C}$ be a class of weak random sources on $\{0,1\}^n$ and let $X \in \mathcal{C}$. Let*

- *$D : \{0,1\}^n \mapsto \{0,1\}^d$ be a zero-error disperser for $\mathcal{C}$.*
- *$E : \{0,1\}^n \times \{0,1\}^d \mapsto \{0,1\}^m$ be a function such that for every $b \in \{0,1\}^m$, there exists $a \in \{0,1\}^d$ and a subset $S \subseteq \mathrm{Supp}(X)$ such that*
  - *S is the support of a distribution in $\mathcal{C}$.*
  - *For all $x \in S$, $E(x,a) = b$.*

Then,

$$F(X) = \{0,1\}^m.$$

We call the function E a *subsource hitter* - for every possible output $b \in \{0,1\}^m$ there is a 'subsource' of X with the support of a distribution in $\mathcal{C}$, namely $(X|X \in S)$, that 'hits' b.

Proof. Given any $b \in \{0,1\}^m$ we need to show there is an $x \in \mathrm{Supp}(X)$ such that $F(x) = E(x, D(x)) = b$. From the guarantee of E we know there is $a \in \{0,1\}^d$ and $S \subseteq \mathrm{Supp}(X)$ such that for every $x \in S$, $E(x,a) = b$. Since S is a support of a distribution in $\mathcal{C}$, from the guarantee of D there must be $x \in S$ such that $D(x) = a$. For this x,

$$F(x) = E(x, D(x)) = E(x,a) = b.$$

□

In Chapter 2 we construct a subsource hitter for 2-independent sources of min-entropy k with output length $\Omega(k)$. Using the Composition Theorem for zero-error dispersers, this enables increasing the output length of the zero-error disperser of [4] for 2-independent sources of min-entropy $k = \delta \cdot n$ for constant δ from a constant number of bits to $\Omega(k)$.

We get similar improvements for zero-error dispersers for bit-fixing sources and affine sources over large fields.

1.4.5 What Else Is There in This Book?

In this introduction we chose to concentrate on the recycling paradigm and showed the composition theorems based on it. A large part of this book, which we did not address here, is devoted to constructing the components that are 'plugged into' the composition theorems. The one work which we did not address in this introduction is described in Chapter 4, where we construct deterministic extractors for 'low-degree polynomial sources'. In that work we do not use the recycling paradigm.

Chapter 2

Deterministic Extractors for Bit-Fixing Sources by Obtaining an Independent Seed

Summary

An (n, k)-bit-fixing source is a distribution X over $\{0,1\}^n$ such that there is a subset of k variables in $X_1, \ldots, X_n$ which are uniformly distributed and independent of each other, and the remaining $n-k$ variables are fixed. A deterministic bit-fixing source extractor is a function $E : \{0,1\}^n \to \{0,1\}^m$ which on an arbitrary (n, k)-bit-fixing source outputs m bits that are statistically-close to uniform. Prior to our work, Kamp and Zuckerman [44th FOCS, 2003] gave a construction of a deterministic bit-fixing source extractor that extracts $\Omega(k^2/n)$ bits and requires $k > \sqrt{n}$.

In this chapter we give constructions of deterministic bit-fixing source extractors that extract $(1 - o(1))k$ bits whenever $k > (\log n)^c$ for some universal constant $c > 0$. Thus, our constructions extract almost all the randomness from bit-fixing sources and work even when k is small. For $k \gg \sqrt{n}$ the extracted bits have statistical distance $2^{-n^{\Omega(1)}}$ from uniform, and for $k \le \sqrt{n}$ the extracted bits have statistical distance $k^{-\Omega(1)}$ from uniform.

Our technique gives a general method to transform deterministic bit-fixing source extractors that extract few bits into extractors which extract almost all the bits. This work is the first to use the 'recycling paradigm' as described in the introduction. The description of it here is different and perhaps more cumbersome, as the one given in the introduction was only realized in hindsight.

This chapter is based on [26].

A. Gabizon, *Deterministic Extraction from Weak Random Sources*, Monographs in Theoretical Computer Science. An EATCS Series, DOI 10.1007/978-3-642-14903-0_2, © Springer-Verlag Berlin Heidelberg 2011

2.1 Introduction

2.1.1 Bit-Fixing Sources

In this chapter we concentrate on the family of "bit-fixing sources" introduced by Chor et al. [15]. A distribution X over $\{0,1\}^n$ is a bit-fixing source if there is a subset $S \subseteq \{1,\ldots,n\}$ of "good indices" such that the bits X_i for $i \in S$ are independent fair coins and the rest of the bits are fixed.[1]

Definition 2.1 (bit-fixing sources and extractors). *A distribution X over $\{0,1\}^n$ is an (n,k)-*bit-fixing source *if there exists a subset $S = \{i_1,\ldots,i_k\} \subseteq \{1,\ldots,n\}$ such that $X_{i_1}, X_{i_2}, \ldots, X_{i_k}$ is uniformly distributed over $\{0,1\}^k$ and for every $i \notin S$, X_i is constant.*

A function $E : \{0,1\}^n \to \{0,1\}^m$ is a deterministic (k,ϵ)-bit-fixing source extractor if it is a deterministic ϵ-extractor (as defined in the first chapter) for all (n,k)-bit-fixing sources.

One of the motivations given in the literature for studying deterministic bit-fixing source extractors is that they are helpful in cryptographic scenarios in which an adversary learns (or alters) $n-k$ bits of an n-bit long secret key [15]. Loosely speaking, one wants cryptographic protocols to remain secure even in the presence of such adversaries. Various models for such "exposure resilient cryptography" were studied [57, 11, 12, 20]. The reader is referred to [18] for a comprehensive treatment of "exposure resilient cryptography" and its relation to deterministic bit-fixing source extractors.

Every (n,k)-bit-fixing source "contains" k "bits of randomness". It follows that any deterministic (k,ϵ)-bit-fixing source extractor with $\epsilon < 1/2$ can extract at most k bits. The function $E(x) = \oplus_{1 \le i \le n} x_i$ is a deterministic $(k,0)$-bit-fixing source extractor which extracts one bit for any $k \ge 1$. Chor et al. [15] concentrated on deterministic "errorless" extractors (that is, deterministic extractors in which $\epsilon = 0$). They show that such extractors cannot extract even two bits when $k < n/3$. They also give some constructions of deterministic errorless extractors for large k.

Our focus is on extractors with error $\epsilon > 0$ (which allows extracting many bits for many choices of k). A probabilistic argument shows the existence of a deterministic (k,ϵ)-bit-fixing source extractor that extracts $m = k - O(\log(n/\epsilon))$ bits for any choice of k and ϵ. Thus, it is natural to try and achieve such parameters by explicit constructions.

In a previous work, Kamp and Zuckerman [37] constructed explicit deterministic (k,ϵ)-bit-fixing source extractors that extract $m = \eta k^2/n$ bits for some constant $0 < \eta < 1$ with $\epsilon = 2^{-\Omega(k^2/n)}$. They pose the open problem to

[1] We remark that such sources are often referred to as "oblivious bit-fixing sources" to differentiate them from other types of "non-oblivious" bit-fixing sources in which the bits outside of S may depend on the bits in S (cf. [7]). In this chapter we are only concerned with the "oblivious case".

extract more bits from such sources. Note that the extractor of Kamp and Zuckerman is inferior to the nonexplicit extractor in two respects:

- It only works when $k > \sqrt{n}$.
- Even when $k > \sqrt{n}$ the extractor may extract only a small fraction of the randomness. For example, if $k = n^{1/2+\alpha}$ for some $0 < \alpha < 1/2$ the extractor only extracts $m = \eta n^{2\alpha}$ bits.

2.1.2 Our Results

We give two constructions of deterministic bit-fixing source extractors that extract $m = (1 - o(1))k$ bits from (n, k)-bit-fixing sources. Our first construction is for the case of $k \gg \sqrt{n}$.

Theorem 2.1. *For every constant $0 < \gamma < 1/2$ there exists an integer n' (depending on γ) such that for any $n > n'$ and any k, there is an explicit deterministic (k, ϵ)-bit-fixing source extractor $E : \{0,1\}^n \to \{0,1\}^m$ where $m = k - n^{1/2+\gamma}$ and $\epsilon = 2^{-\Omega(n^\gamma)}$.*

Consider $k = n^{1/2+\alpha}$ for some constant $0 < \alpha < 1/2$. We can choose any $\gamma < \alpha$ and extract $m = n^{1/2+\alpha} - n^{1/2+\gamma}$ bits whereas the construction of [37] only extracts $m = O(n^{2\alpha})$ bits. For this choice of parameters we achieve error $\epsilon = 2^{-\Omega(n^\gamma)}$ whereas [37] achieves a slightly smaller error $\epsilon = 2^{-\Omega(n^{2\alpha})}$. We remark that this comes close to the parameters achieved by the nonexplicit construction, which can extract $m = n^{1/2+\alpha} - n^{1/2+\gamma}$ with error $\epsilon = 2^{-\Omega(n^{1/2+\gamma})}$.

Our second construction works for any $k > (\log n)^c$ for some universal constant c. However, the error in this construction is larger.

Theorem 2.2. *There exist constants $c > 0$ and $0 < \mu, \nu < 1$ such that for any large enough n and any $k \geq \log^c n$, there is an explicit deterministic (k, ϵ)-bit-fixing source extractor $E : \{0,1\}^n \to \{0,1\}^m$ where $m = k - O(k^\nu)$ and $\epsilon = O(k^{-\mu})$.*

We remark that using the technique of [37] one can achieve much smaller error ($\epsilon = 2^{-\sqrt{k}}$) at the cost of extracting very few bits ($m = \Omega(\log k)$). The precise details are given in Theorem 2.5.

2.1.3 Overview of Techniques

We develop a general technique that transforms any deterministic bit-fixing source extractor that extracts only very few bits into one that extracts almost all of the randomness in the source. This transformation makes use of "seeded extractors".

Seeded randomness extractors

A seeded randomness extractor is a function which receives two inputs: In addition to a sample from a source X, a seeded extractor also receives a short "seed" Y of few uniformly distributed bits. Loosely speaking, the extractor is required to output many more random bits than the number of bits "invested" as a seed.

Definition 2.2 (seeded extractors). *Let $\mathcal{C}$ be a class of distributions on $\{0,1\}^n$. A function $E : \{0,1\}^n \times \{0,1\}^d \to \{0,1\}^m$ is a* seeded ϵ-extractor *for $\mathcal{C}$ if for every source X in $\mathcal{C}$ the distribution $E(X, Y)$ (obtained by sampling x from X and a uniform $y \in \{0,1\}^d$ and computing $E(x, y)$) is ϵ-close to the uniform distribution on m bit strings.*

A long line of research focuses on constructing such seeded extractors with as short as possible seed length that extract as many as possible bits from the most general family of sources that allow randomness extraction: The class of sources with high min-entropy.

Definition 2.3 (seeded extractors for high min-entropy sources). *The min-entropy of a distribution X over $\{0,1\}^n$ is $H_\infty(X) = min_{x \in \{0,1\}^n} \log_2 (1/\Pr(x))$. A function $E : \{0,1\}^n \times \{0,1\}^d \to \{0,1\}^m$ is a* (k, ϵ)-extractor *if it is a seeded ϵ-extractor for the class of all sources X with $H_\infty(X) \geq k$.*

There are explicit constructions of (k, ϵ)-extractors that use a seed of length polylog(n/ϵ) to extract k random bits. The reader is referred to [62] for a detailed survey on various constructions of seeded extractors.

Our goal is to construct *deterministic* bit-fixing source extractors. Nevertheless, in the next definition we introduce the concept of a *seeded* bit-fixing source extractor. We use such extractors as a component in our construction of deterministic bit-fixing source extractors.

Definition 2.4 (seeded extractors for bit-fixing sources). *A function $E : \{0,1\}^n \times \{0,1\}^d \to \{0,1\}^m$ is a* seeded (k, ϵ)-bit-fixing source extractor *if it is a seeded ϵ-extractor for the class of all (n, k)-bit-fixing sources.*

Seed obtainers

There is a very natural way to try to transform a deterministic bit-fixing source extractor that extracts few (say polylog(n)) bits into one that extracts many bits: First, run the deterministic bit-fixing source extractor to extract a few bits from the source, and then use these bits as a seed to a seeded extractor that extracts all the bits from the source. The obvious difficulty with this approach is that typically the output of the first extractor is *correlated* with the imperfect random source. Seeded extractors are only guaranteed to work when their seed is *independent* from the random source. To overcome this difficulty we introduce a new object we call a *"seed obtainer"*.

Loosely speaking, a seed obtainer is a function F that, given an (n, k)-bit-fixing source X, outputs two strings X' and Y with the following properties:

- X' is an (n, k')-bit-fixing source with $k' \approx k$ good bits.
- Y is a short string that is almost uniformly distributed.
- X' and Y are almost independent.

The precise definition is slightly more technical and is given in Definition 2.11. Note that a seed obtainer reduces the task of constructing *deterministic* extractors to that of constructing *seeded* extractors: Given a bit-fixing source X, one first runs the seed obtainer to obtain X' and a short Y, and then uses Y as a seed to a seeded extractor that extracts all the randomness from X'. (In fact, it is even sufficient to construct seeded extractors for bit-fixing sources.)

Constructing seed obtainers

Note that every seed obtainer $F(X) = (X', Y)$ "contains" a deterministic bit-fixing source extractor by setting $E(X) = Y$. We show how to transform any deterministic bit-fixing source extractor into a seed obtainer. In this transformation the length of the "generated seed" Y is roughly the length of the output of the original extractor.

It is helpful to explain the intuition behind this transformation when applied to a specific deterministic bit-fixing source extractor. Consider the "xor-extractor" $E(x) = \oplus_{1 \le i \le n} x_i$. Let X be some (n, k)-bit-fixing source, and let $Z = E(X)$. Note that the output bit Z is indeed very correlated with the input X. Nevertheless, suppose that we somehow obtain a random small subset of the indices of X. It is expected that the set contains a small fraction of the good bits. Let X' be the string that remains after "removing" the indices in the sampled set. The important observation is that X' is a bit-fixing source that is *independent* from the output Z. It turns out that the same phenomenon happens for every deterministic bit-fixing source extractor $E(X)$. However, it is not clear how to use this idea as we don't have additional random bits to perform the aforementioned sampling of a random set. Surprisingly, we show how to use the bits extracted by the extractor E to perform this sampling.

Following this intuition, given an extractor $E(X)$ which extracts an m bit string Z, we partition Z into two parts Y and W. We then use W as a seed to a randomness efficient method of "sampling" a small subset T of $\{1, \ldots, n\}$. The first output of the seed obtainer X' is given by "removing" the sampled indices from X. More formally, X' is the string X restricted to the indices outside of T. The second output is Y (the other part of the output of the extractor E).

The intuition is that if T was a size n/r uniformly distributed subset of $\{1, \ldots, n\}$ then it is expected to hit approximately k/r good bits from the source. Thus, $k - k/r$ good bits remain in X'. We will require that the extractor E extracts randomness from $(n, k/r)$-bit-fixing sources. Loosely speaking, we can hope that E will extract its output from X_T (the string obtained by restricting X to the indices of T). Thus, its output will be independent from X' (the string obtained by removing X_T).

Note that the intuition above is far from being precise. The set T is sampled using random bits W that are extracted from the source X, and thus T depends on X, whereas, the intuition corresponds to the case where T is independent from X. The precise argument appears in Section 2.3. We remark that the analysis requires that the extractor E have error ϵ that is smaller than $2^{-|W|}$ (where $|W|$ is the number of bits used by the sampling method).

A deterministic extractor for large k (i.e., $k \gg \sqrt{n}$)

Our first construction builds on the deterministic bit-fixing source extractor of Kamp and Zuckerman [37] that works for $k > \sqrt{n}$ and extracts at least $\Omega(k^2/n)$ bits from the source. We first transform this extractor into a seed obtainer F. Next, we run the seed obtainer F on the input source to generate a bit-fixing source X' and a seed Y. Finally, we extract all the randomness in X' by running a seeded extractor on X' using Y as seed.

A deterministic extractor for small k (i.e., $k < \sqrt{n}$)

In order to use our technique for $k < \sqrt{n}$ we need to start with some deterministic bit-fixing source extractor that works when $k < \sqrt{n}$ and extracts a small number of bits. Our first observation is that methods similar to the ones of Kamp and Zuckerman [37] can be applied when $k < \sqrt{n}$ but only give deterministic bit-fixing source extractors that extract very few bits (i.e., $\Omega(\log k)$ bits)[2].

Deterministic extractors that extract $\Omega(\log k)$ bits Kamp and Zuckerman [37] consider the distribution obtained by using a bit-fixing source $X = (X_1, \ldots, X_n)$ to perform a random walk on a d-regular graph. (They consider a more general model of bit-fixing sources in which every symbol X_i ranges over an alphabet of size d). The walk starts from some fixed vertex in the graph, and at step i X_i is used to select a neighbor of the current vertex. They show that the distribution over the vertices converges to the uniform distribution at a rate which depends on k and the "spectral gap" of the graph. It is known that 2-regular graphs cannot have a small "spectral gap". Indeed, this is why Kamp and Zuckerman consider alphabet size $d > 2$, which allows using d-regular expander graphs that have small spectral gap. Nevertheless,

[2]This was observed independently by Lipton and Vishnoi [40].

using their technique choosing the graph to be a short cycle of length $k^{1/4}$ produces an extractor construction which extracts $\log(k^{1/4}) = \Omega(\log k)$ bits.[3]

A seeded extractor for bit-fixing sources with seed length $O(\log\log n)$ Converting the deterministic bit-fixing source extractor above into a seed obtainer we "obtain" an $\Omega(\log k)$ bit seed. This allows us to use a seeded extractor with seed length $d = \Omega(\log k)$. However, $d < \log n$ and by a lower bound of [48, 49] the class of high min-entropy sources does not have seeded extractors with seed $d < \log n$. To bypass this problem we construct a *seeded* extractor for *bit-fixing sources* with seed length $O(\log\log n)$. Note that the aforementioned deterministic extractor extracts these many bits as long as $k > \log^c n$ for some constant c (when $\Omega(\log k) \geq O(\log\log n)$).

The seeded extractor uses its seed to randomly partition the indices $\{1, \ldots, n\}$ into r sets $T_1, \ldots, T_r$ (for r equal, say to $\log^4 n$), with the property that with high probability each one of these sets contains at least one good bit. We elaborate on this partitioning method later on. We then output r bits, where the i'th bit is given by $\oplus_{j \in T_i} x_j$.

By combining the seed obtainer with the seeded bit-fixing source extractor we obtain a deterministic bit-fixing source extractor which extracts $r = \log^4 n$ bits. To extract more bits, we convert this deterministic extractor into a seed obtainer. At this point we obtain a seed of length $\log^4 n$ and can afford using a seeded extractor which extracts all the remaining randomness.

Sampling and partitioning with only $O(\log\log n)$ **random bits** We now explain how to use $O(\log\log n)$ random bits to partition the indices $\{1, \ldots, n\}$ into $r = \operatorname{poly}\log n$ sets $T_1, \ldots, T_r$ such that for any set $S \subseteq \{1, \ldots, n\}$ of size k, with high probability (probability at least $1 - O(1/\log n)$) all sets $T_1, \ldots, T_r$ contain approximately k/r indices from S.

Suppose we could afford using many random bits. A natural solution is to choose n random variables $V_1, \ldots, V_n \in \{1, \ldots, r\}$ and have T_j be the set of indices i such that $V_i = j$. We expect k/r bits to fall in each T_j and by a union bound one can show that with high probability all sets $T_1, \ldots, T_r$ have a number of indices from S that is close to the expected value.

To reduce the number of random bits we derandomize the construction above and use random variables V_i which are ϵ-close to being pairwise independent (for $\epsilon = 1/\log^a n$ for some sufficiently large constant a). Such variables can be constructed using only $O(\log\log n)$ random bits [44, 2, 23] and suffice to guarantee the required properties.

The same technique also gives us a method for sampling a set T of indices in $\{1, \ldots, n\}$ (which we require in our construction of seed obtainers). We simply take the first set T_1. This sampling method uses only $O(\log\log n)$ random bits and thus we can afford it when transforming our deterministic

[3] In fact, a similar idea is used in [37] in order to reduce the case of large d to the case of $d = 2$.

extractor into a seed obtainer. (Recall that our transformation uses part of the output of the deterministic extractor for sampling a subset of the indices). We remark that this sampling technique was used previously by Reingold et al. [56] as a component in a construction of seeded extractors.

2.1.4 Outline

In Section 2.2 we define the notations used in this chapter. In Section 2.3 we introduce the concept of seed obtainers and show how to construct them from deterministic bit-fixing source extractors and "averaging samplers". In Section 2.4 we observe that the technique of [37] can be used to extract a few bits even when k is small. In Section 2.5 we give constructions for averaging samplers. In Section 2.6 we give a construction of a seeded bit-fixing source extractor that makes use of the sampling techniques of Section 2.5. In Section 2.7 we plug all the components together and prove our main theorems. Finally, in Section 2.8 we give some open problems.

2.2 Preliminaries

Notations We use $[n]$ to denote the set $\{1,\ldots,n\}$. We use $P(S)$ to denote the set of subsets of a given set S. We use U_n to denote the uniform distribution over n bits. Given a distribution A we use $w \leftarrow A$ to denote the experiment in which w is chosen randomly according to A. Given a string $x \in \{0,1\}^n$ and a set $S \subseteq [n]$ we use x_S to denote the string obtained by restricting x to the indices in S. We denote the length of a string x by $|x|$. Logarithms will always be taken with base 2.

Asymptotic notation As this chapter has many parameters, we now explain exactly what we mean when using $O(\cdot)$ and $\Omega(\cdot)$ in a statement involving many parameters. We use the Ω and O signs only to denote absolute constants (i.e., constants not dependent on any parameters even if these parameters are considered constants). Furthermore, when writing, for example, $f(n) = O(g(n))$, we always explicitly mention the conditions on n (and maybe other parameters) for which the statement holds.

2.2.1 Averaging Samplers

A sampler is a procedure which, given a short seed, generates a subset $T \subseteq [n]$ such that for every set $S \subseteq [n]$, $|S \cap T|$ is with high probability "close to the expected size".

Definition 2.5. *An $(n, k, k_{min}, k_{max}, \delta)$-sampler $Samp : \{0,1\}^t \to P([n])$ is a function such that for any $S \subseteq [n]$ such that $|S| = k$,*

$$\Pr_{w \leftarrow U_t}(k_{min} \leq |Samp(w) \cap S| \leq k_{max}) \geq 1 - \delta.$$

The definition above is nonstandard in several respects. In the more common definition (c.f. [30]), a sampler is required to work for sets of arbitrary size. In the definition above (which is similar in spirit to the one in [69]), the sampler is only required to work against sets of size k and the bounds k_{min}, k_{max} are allowed to depend on k. Furthermore, we require that the sampler have "distinct samples" as we do not allow T to be a multi-set.[4]

We will use samplers to "partition" bit-fixing sources. Note that in the case of an (n, k)-bit-fixing source, *Samp* returns a subset of indices such that, with high probability, the number of good bits in the subset is between k_{min} and k_{max}.

2.2.2 Probability Distributions

Some of the proofs in this chapter require careful manipulations of probability distributions. We use the following notation. We use U_m to denote the uniform distribution on m bit strings. We denote the probability of an event B under a probability distribution P by $\Pr_P[B]$. A random variable R that takes values in U is a function $R : \Omega \to U$ (where Ω is a probability space). We sometimes refer to R as a probability distribution over U (the distribution of the output of R). For example, given a random variable R and a distribution P we sometimes write "$R = P$", and this means that the distribution of the output of R is equal to P. Given two random variables R_1, R_2 over the same probability space Ω we use (R_1, R_2) to denote the random variable induced by the function $(R_1, R_2)(\omega) = (R_1(\omega), R_2(\omega))$. Given two probability distributions P_1, P_2 over domains Ω_1, Ω_2, we define $P_1 \otimes P_2$ to be the product distribution of P_1 and P_2, defined over the domain $\Omega_1 \times \Omega_2$.

Definition 2.6 (conditioning distributions and random variables). *Given a probability distribution P over some domain U and an event $A \subseteq U$ such that $\Pr_P[A] > 0$ we define a distribution $(P|A)$ over U as follows: Given an event $B \subseteq U$, $\Pr_{(P|A)}(B) = \Pr_P[B|A] = \frac{\Pr_P[A \cap B]}{\Pr_P[A]}$.*

We extend this definition to random variables $R : \Omega \to U$. Given an event $A \subseteq \Omega$ we define $(R|A)$ to be the probability distribution over U given by $\Pr_{(R|A)}[B] = \Pr_R[R \in B|A]$.

We also need the notion of convex combination of distributions.

Definition 2.7 (convex combination of distributions). *Given distributions $P_1, \ldots, P_t$ over U and coefficients $\alpha_1, \ldots, \alpha_t \geq 0$ such that $\sum_{1 \leq i \leq t} \alpha_i = 1$, we*

[4] We remark that some of the "standard techniques" for constructing averaging samplers (such as taking a walk on an expander graph or using a randomness extractor) perform poorly in this setup, and do not work when $k < \sqrt{n}$ (even if T is allowed to be a multi-set). This happens because in order to even hit a set S of size k, these techniques require sampling a (multi-)set T of size larger than $(n/k)^2$, which is larger than n for $k < \sqrt{n}$. In contrast, note that a completely random set of size roughly n/k will hit a fixed set S of small size with high probability.

define the distribution $P = \sum_{1 \le i \le t} \alpha_i P_i$ *as follows: Given an event* $B \subseteq U$, $\Pr_P[B] = \sum_{1 \le i \le t} \alpha_i \Pr_{P_i}[B]$.

We also need the following technical lemmas.

Lemma 2.8. *Let* X, Y *and* V *be distributions over* $\{0,1\}^n$ *such that* X *is* ϵ*-close to* U_n *and* $Y = \delta \cdot V + (1-\delta) \cdot X$. *Then* Y *is* $(2\delta + \epsilon)$*-close to* U_n.

Proof. Let $B \subseteq \{0,1\}^n$ be some event.

$$\begin{aligned} |\Pr_Y(B) - \Pr_{U_n}(B)| &= |\delta \Pr_V(B) + (1-\delta)\Pr_X(B) - \Pr_{U_n}(B)| \\ &\le 2\delta + |\Pr_X(B) - \Pr_{U_n}(B)| \le 2\delta + \epsilon. \end{aligned}$$

□

Lemma 2.9. *Let* (A, B) *be a random variable that takes values in* $\{0,1\}^u \times \{0,1\}^v$ *and suppose that there exists some distribution* P *over* $\{0,1\}^v$ *such that for every* $a \in \{0,1\}^u$ *with* $\Pr[A = a] > 0$ *the distribution* $(B|A = a)$ *is* ϵ*-close to* P. *Then* (A, B) *is* ϵ*-close to* $(A \otimes P)$.

Proof.

$$\frac{1}{2} \cdot \sum_{a,b} |\Pr[(A,B) = (a,b)] - \Pr_{A \otimes P}[a,b]|$$

$$= \frac{1}{2} \cdot \sum_{a,b} |\Pr[A = a]\Pr[B = b|A = a] - \Pr[A = a]\Pr_P[b]|$$

$$\le \frac{1}{2} \cdot \sum_a \Pr[A = a] \sum_b |\Pr[B = b|A = a] - \Pr_P[b]| \le \epsilon/2.$$

□

Lemma 2.10. *Let* (A, B) *be a random variable that takes values in* $\{0,1\}^u \times \{0,1\}^v$ *which is* ϵ*-close to* $(A' \otimes U_v)$*; then, for every* $b \in \{0,1\}^v$ *the distribution* $(A|B = b)$ *is* $(\epsilon \cdot 2^{v+1})$*-close to* A'.

Proof. Assume for the purpose of contradiction that there exists some $b^* \in \{0,1\}^v$ such that the distribution $(A|B = b^*)$ is not α-close to A' for $\alpha = \epsilon \cdot 2^{v+1}$. Then there is an event D such that

$$|\Pr_{(A|B=b^*)}[D] - \Pr_{A'}[D]| > \alpha.$$

By complementing D if necessary, we can w.l.o.g. remove the absolute value from the inequality above. We define an event D' over $\{0,1\}^u \times \{0,1\}^v$. The event $D' = \{(a,b) | b = b^*,\ a \in D\}$. We have that

$$\Pr_{(A', U_v)}[D'] = \Pr_{A'}[D] \cdot 2^{-v}.$$

And similarly,

$$\Pr_{(A,B)}[D'] = \Pr_{(A|B=b^*)}[D] \Pr_B[B = b^*].$$

We know that B is ϵ-close to U_v, and therefore $\Pr_B[B = b^*] \geq 2^{-v} - \epsilon$. Thus,

$$\Pr_{(A,B)}[D'] - \Pr_{(A',U_v)}[D'] = \Pr_{(A|B=b^*)}[D] \Pr_B[B = b^*] - \Pr_{A'}[D] \cdot 2^{-v}$$

$$\geq \Pr_{(A|B=b^*)}[D](2^{-v} - \epsilon) - \Pr_{A'}[D] \cdot 2^{-v} \geq 2^{-v}[\Pr_{(A|B=b^*)}[D] - \Pr_{A'}[D]] - \epsilon.$$

By our assumption the expression in square brackets is at least α, and thus,

$$> 2^{-v}\alpha - \epsilon = \epsilon.$$

Thus, we get a contradiction. □

2.3 Obtaining an Independent Seed

2.3.1 Seed Obtainers and Their Application

One of the natural ways to try and extract *many* bits from imperfect random sources is to first run a "weak extractor" which extracts few bits from the input distribution and then use these few bits as a seed to a second extractor which extracts more bits. The obvious difficulty with this approach is that typically the output of the first extractor is *correlated* with the imperfect random source and it is not clear how to use it. (Seeded extractors are only guaranteed to work when the seed is *independent* from the random source). In the next definition we introduce the concept of a *"seed obtainer"* that overcomes this difficulty. Loosely speaking, a seed obtainer is a deterministic function which given a bit-fixing source X outputs a new bit-fixing source X' (with roughly the same randomness) together with a short random seed Y which is *independent* from X'. Thus, the seed Y can later be used to extract randomness from X' using a seeded extractor.

Definition 2.11 (seed obtainer). *A function $F : \{0,1\}^n \to \{0,1\}^n \times \{0,1\}^d$ is a (k, k', ρ)-seed obtainer if for every (n, k)-bit-fixing source X, the distribution $R = F(X)$ can be expressed as a convex combination of distributions $R = \eta Q + \sum_a \alpha_a R_a$ (here the coefficients η and α_a are nonnegative and $\eta + \sum_a \alpha_a = 1$) such that $\eta \leq \rho$ and for every a there exists an (n, k')-bit-fixing source Z_a such that R_a is ρ-close to $Z_a \otimes U_d$.*

It follows that given a seed obtainer one can use a *seeded extractor* for bit-fixing sources to construct a *deterministic* (i.e., seedless) extractor for bit-fixing sources.

Theorem 2.3. *Let $F : \{0,1\}^n \to \{0,1\}^n \times \{0,1\}^d$ be a (k, k', ρ)-seed obtainer. Let $E_1 : \{0,1\}^n \times \{0,1\}^d \to \{0,1\}^m$ be a seeded (k', ϵ)-bit-fixing source extractor. Then $E : \{0,1\}^n \to \{0,1\}^m$ defined by $E(x) = E_1(F(x))$ is a deterministic $(k, \epsilon + 3\rho)$-bit-fixing source extractor*

Proof. By the definition of a seed obtainer we have that $E(X) = \eta E_1(Q) + \sum_a \alpha_a E_1(R_a)$ for some $\eta \leq \rho$. For each a we have that $E_1(R_a)$ is $(\epsilon + \rho)$-close to U_m. It follows that $E(X)$ is $(\epsilon + \rho)$-close to $\eta E_1(Q) + (1-\eta)U_m$ and therefore by Lemma 2.8 we have that $E(X)$ is $(2\eta + \epsilon + \rho)$-close to uniform. The lemma follows because $2\eta + \epsilon + \rho \leq \epsilon + 3\rho$. □

2.3.2 Constructing Seed Obtainers

Note that every seed obtainer "contains" a deterministic extractor for bit-fixing sources. More precisely, given a seed obtainer $F(x) = (x', y)$ the function $E(x) = y$ is a deterministic extractor for bit-fixing sources. We now show how to convert any deterministic bit-fixing source extractor with sufficiently small error into a seed obtainer.

Our construction appears in Figure 2.1. In words, given x, the seed obtainer first computes $E(x)$. It uses a part of $E(x)$ as the second output y and another part to sample a substring of x. It obtains the first output x' by erasing the sampled substring from x. We now state the main theorem of this section.

Theorem 2.4 (construction of seed obtainers). *For every n and $k < n$, Let Samp and E be as in Figure 2.1 (that is, $Samp : \{0,1\}^t \to P([n])$ is*

Figure 2.1: A seed obtainer for (n, k)-bit-fixing sources

Ingredients:

- An $(n, k, k_{min}, k_{max}, \delta)$-sampler $Samp : \{0,1\}^t \to P([n])$.
- A deterministic (k_{min}, ϵ)-bit-fixing source extractor $E : \{0,1\}^n \to \{0,1\}^m$ with $m > t$.

Result: A (k, k', ρ)-seed obtainer $F : \{0,1\}^n \to \{0,1\}^n \times \{0,1\}^{m-t}$ with $k' = k - k_{max}$ and $\rho = \max(\epsilon + \delta, \epsilon \cdot 2^{t+1})$.

The construction of F:

- Given $x \in \{0,1\}^n$ compute $E(x)$ and let $E_1(x)$ denote the first t bits of $E(x)$ and $E_2(x)$ denote the remaining $m - t$ bits.
- Let $T = Samp(E_1(x))$.
- Let $x' = x_{[n]\setminus T}$. If $|x'| < n$ we pad it with zeroes to get an n-bit long string.
- Let $y = E_2(x)$, Output x', y.

an $(n, k, k_{min}, k_{max}, \delta)$-sampler and $E : \{0,1\}^n \to \{0,1\}^m$ is a deterministic (k_{min}, ϵ)-bit-fixing source extractor). Then, $F : \{0,1\}^n \to \{0,1\}^n \times \{0,1\}^d$ defined in Figure 2.1 is a (k, k', ρ)-seed obtainer for $d = m - t$, $k' = k - k_{max}$ and $\rho = \max(\epsilon + \delta, \epsilon \cdot 2^{t+1})$.

Proof of Theorem 2.4

In this section we prove Theorem 2.4. Let E be a bit-fixing source extractor and $Samp$ be a sampler which satisfy the requirements in Theorem 2.4. Let X be some (n, k)-bit-fixing source and let $S \subseteq [n]$ be the set of k good indices for X. We will use capital letters to denote the random variables which come up in the construction. We split $E(X)$ into two parts $(E_1(X), E_2(X)) \in \{0,1\}^t \times \{0,1\}^{m-t}$. For a string $a \in \{0,1\}^t$ we use T_a to denote $Samp(a)$ and T'_a to denote $[n] \setminus Samp(a)$. Given a string $x \in \{0,1\}^n$, we use x_a to denote x_{T_a} and x'_a to denote the n bit string obtained by padding $x_{T'_a}$ to length n. Let $X' = X'_{E_1(X)}$ and $Y = E_2(X)$. Our goal is to show that the pair (X', Y) is close to a convex combination of pairs of distributions where the first component is a bit-fixing source and the second is independent and uniformly distributed.

Definition 2.12. *We say that a string $a \in \{0,1\}^t$* correctly splits *X if $k_{min} \leq |S \cap T_a| \leq k_{max}$.*

Note that by the properties of the sampler, almost all strings a correctly split X. We start by showing that for every *fixed* a which correctly splits X the variables X'_a and $E(X)$ are essentially independent. Loosely speaking this happens because we can argue that there are enough good bits in X_a and therefore the extractor can extract randomness from X_a which is independent of the randomness in X'_a.

Lemma 2.13. *For every fixed $a \in \{0,1\}^t$ which correctly splits X the pair of random variables $(X'_a, E(X))$ is ϵ-close to the pair $(X'_a \otimes U_m)$.*

Proof. Let $\ell = |Samp(a)|$. Given a string $\sigma \in \{0,1\}^\ell$ and a string $\sigma' \in \{0,1\}^{n-\ell}$ we define $[\sigma; \sigma']$ to be the n bit string obtained by placing σ in the indices of T_a and σ' in the indices of T'_a. More formally, we denote the ℓ indices of T_a by $i_1 < i_2 < \cdots < i_\ell$ and the $n - \ell$ indices of T'_a by $i'_1 < i'_2 < \cdots < i'_{n-\ell}$. Given an $i \in T_a$ we define $index(i)$ to be the index j such that $i_j = i$, and equivalently given $i \in T'_a$ we define $index'(i)$ to be the index j such that $i'_j = i$. The string $[\sigma; \sigma'] \in \{0,1\}^n$ is defined as follows:

$$[\sigma; \sigma']_i = \begin{cases} \sigma_{index(i)} & i \in T_a \\ \sigma'_{index'(i)} & i \in T'_a \end{cases}$$

Note that in this notation $X = [X_a; X'_a]$. We are interested in the distribution of the random variable $(X'_a, E(X)) = (X'_a, E([X_a; X'_a]))$. For every

$b \in \{0,1\}^{n-\ell}$ we consider the event $\{X'_a = b\}$. Fix some $b \in \{0,1\}^{n-\ell}$ such that $\Pr[X'_a = b] > 0$. The distribution

$$(E(X)|X'_a = b) = (E([X_a; X'_a])|X'_a = b) = E([X_a; b])$$

where the last equality follows because X_a and X'_a are independent and therefore X_a is not affected by fixing X'_a. Note that as a correctly splits X, the distribution $[X_a; b]$ is a bit-fixing source with at least k_{min} "good" bits. We conclude that for every $b \in \{0,1\}^{n-\ell}$ such that $\Pr[X'_a = b] > 0$ the distribution $(E(X)|X'_a = b)$ is ϵ-close to uniform. We now apply Lemma 2.9 with $A = X'_a$ and $B = E(X)$ and conclude that the pair $(X'_a, E(X))$ is ϵ-close to $(X'_a \otimes U_m)$. □

We now argue that if ϵ is small enough then the pair $(X'_a, E_2(X))$ is essentially independent even when conditioning the probability space on the event $\{E_1(X) = a\}$.

Lemma 2.14. *For every fixed $a \in \{0,1\}^t$ that correctly splits X, the distribution $((X'_a, E_2(X))|E_1(X) = a)$ is $\epsilon \cdot 2^{t+1}$-close to $(X'_a \otimes U_{m-t})$.*

Proof. First note that the statement is meaningless unless $\epsilon < 2^{-t}$ we will assume w.l.o.g. that this is the case and then for every fixed $a \in \{0,1\}^t$ the event $\{E_1(X) = a\}$ occurs with nonzero probability as $E_1(X)$ is ϵ-close to uniform over $\{0,1\}^t$. The lemma will follow as a straightforward application of Lemma 2.10. We set $A = (X'_a, E_2(X))$, $B = E_1(X)$ and $A' = (X'_a, U_{m-t})$. We indeed have that (A, B) is ϵ-close to (A', U_t) and the lemma follows. □

We are now ready to prove Theorem 2.4.

Proof. (of Theorem 2.4) By the properties of the extractor we have that $E_1(X)$ is ϵ-close to uniform. It follows (by the properties of the sampler) that the probability that $E_1(X)$ correctly splits X is $1 - \eta$ for some η which satisfies $\eta \le \epsilon + \delta$. We now consider the output random variable $R = (X', E_2(X))$. We need to express this random variable as a convex combination of independent distributions and a small error term. We set Q to be the distribution (R|"$E_1(X)$ doesn't correctly split X"). For every correctly splitting a we set R_a to be the distribution $(R|E_1(X) = a)$ and $\alpha_a = \Pr[E_1(X) = a]$. By our definition we have that indeed $R = \eta Q + \sum_a \alpha_a R_a$. For every a that correctly splits X we have that $R_a = ((X', E_2(X))|E_1(X) = a) = ((X'_{E_1(X)}, E_2(X))|E_1(X) = a) = ((X'_a, E_2(X))|E_1(X) = a)$. By Lemma 2.14 we have that R_a is $\epsilon \cdot 2^{t+1}$-close to $(X'_a \otimes U_{m-t})$. As a correctly splits X we have that X'_a is an $(n, k - k_{max})$-bit-fixing source as required. Thus, we have shown that the distribution R_a is close to a convex combination of pairs of essentially independent distributions where the first is a bit-fixing source and the second is uniform. □

2.4 Extracting a Few Bits for Any k

The deterministic bit-fixing source extractor of Kamp and Zuckerman [37] only works for $k > \sqrt{n}$. However, their technique easily gives a deterministic bit-fixing source extractor that extracts very few bits ($\Omega(\log k)$ bits) from a bit-fixing source with arbitrarily small k. We will later use this extractor to construct a seed obtainer that will enable us to extract many more bits.

Theorem 2.5. *For every $n > k \geq 100$ there is an explicit deterministic $(k, 2^{-\sqrt{k}})$-bit-fixing source extractor $E : \{0,1\}^n \to \{0,1\}^{(\log k)/4}$.*

For the proof, we need the following result, which is a very special case of Lemma 3.3 in [37].

Lemma 2.15. *([37, Lemma 3.3] for $\epsilon = 0$ and $d = 2$). Let the graph G be an odd cycle with M vertices and second eigenvalue λ. Suppose we take a walk on G for n steps, starting from some fixed vertex v with the steps taken according to the symbols from an (n,k)-bit-fixing source X. Let Z be the distribution on the vertices at the end of the walk; then Z is $\left(\frac{1}{2}\lambda^k\sqrt{M}\right)$-close to the uniform distribution on $[M]$.*

To extract a few bits from a bit-fixing source X, we will use the bits of X to conduct a random walk on a small cycle.

Proof. (of Theorem 2.5) We use the source-string to take a walk on a cycle of size $\sqrt[4]{k}$ from a fixed vertex. The second eigenvalue of a d-cycle is $\cos(\frac{\pi}{d})$ ([41, Ex. 11.1]). Using Lemma 2.15, we reach distance $\left(\cos\left(\frac{\pi}{\sqrt[4]{k}}\right)\right)^k k^{1/8}$ from uniform. By the Taylor expansion of cos, for $0 < x < 1$

$$\cos(x) < 1 - \frac{x^2}{2} + \frac{x^4}{24} < 1 - \frac{x^2}{4}$$

Therefore

$$\left(\cos\left(\frac{\pi}{\sqrt[4]{k}}\right)\right)^k < \left(1 - \frac{\pi^2}{4\sqrt{k}}\right)^k$$
$$< \left(e^{-\frac{\pi^2}{4}}\right)^{\sqrt{k}} < 4^{-\sqrt{k}}$$

where the second to last inequality holds because $(1-x) < e^{-x}$ for $0 < x < 1$. Therefore, we reach distance $4^{-\sqrt{k}}k^{1/8} \leq 2^{-\sqrt{k}}$. By outputting the final vertex's name we get $\frac{\log(k)}{4}$ bits with the same distance from uniform. □

2.5 Sampling and Partitioning with a Short Seed

Let $S \subseteq [n]$ be some subset of size k. In this section we show how to use few random bits in order perform two related tasks.

Sampling: Generate a subset $T \subseteq [n]$ such that $|S \cap T|$ is in a prespecified interval $[k_{min}, k_{max}]$ (see Definition 2.5).

Partitioning: Partition $[n]$ into r distinct subsets $T_1, \ldots, T_r$ such that for every $1 \leq i \leq r$, $|S \cap T_i|$ is in a prespecified interval $[k_{min}, k_{max}]$. Needless to say, a partitioning scheme immediately implies a sampling scheme by concentrating on a single T_i.

In this section we present two constructions of such schemes. The first construction is used in our deterministic bit-fixing source extractor for $k > \sqrt{n}$. In this setup we can allow the sampler to use many random bits (say $n^{\Omega(1)}$ bits) and can have error $2^{-n^{\Omega(1)}}$.

Lemma 2.16 (sampling with low error). *Fix any constants $0 < \gamma \leq 1/2$ and $\alpha > 0$. There exists a constant n' depending on α and γ such that for any integers n, k satisfying $n > n'$ and $n^{1/2+\gamma} \leq k \leq n$, there exists an $(n, k, (n^{1/2+\gamma})/6, n^{1/2+\gamma}, 2^{-\Omega(\alpha \cdot n^{\gamma})})$-sampler $Samp : \{0,1\}^t \to P([n])$ where $t = \alpha \cdot n^{2\gamma}$.*

The second construction is used in our deterministic bit-fixing source extractor for small k. For that construction we require schemes that use only $\alpha \log k$ bits for some small constant $\alpha > 0$. The construction of Lemma 2.16 requires at least $\log n > \log k$ bits, which is too much. Instead, we use a different construction which has much larger error (e.g., $k^{-\Omega(1)}$).

Lemma 2.17 (sampling with $O(\log k)$ bits). *Fix any constant $0 < \alpha < 1$. There exist constants $c > 0, 0 < b < 1$ and $1/2 < e < 1$ (all depending on α) such that for any $n \geq 16$ and $k \geq \log^c n$, we obtain an explicit $(n, k, k^e/2, 3 \cdot k^e, O(k^{-b}))$-sampler $Samp : \{0,1\}^t \to P([n])$ where $t = \alpha \cdot \log k$.*

Lemma 2.18 (partitioning with $O(\log k)$ bits). *Fix any constant $0 < \alpha < 1$. There exist constants $c > 0, 0 < b < 1$ and $1/2 < e < 1$ (all depending on α) such that for any $n \geq 16$ and $k \geq \log^c n$, we can use $\alpha \cdot \log k$ random bits to explicitly partition $[n]$ into $m = \Omega(k^b)$ sets $T_1, \ldots, T_m$ such that for any $S \subseteq [n]$ where $|S| = k$*

$$\Pr(\forall i, \quad k^e/2 \leq |T_i \cap S| \leq 3 \cdot k^e) \geq 1 - O(k^{-b}).$$

The first construction is based on "ℓ-wise independence", and the second is based on "almost 2-wise dependence" [44, 2, 23]. Sampling techniques based on ℓ-wise independence were first suggested by Bellare and Rompel [6]. However, this technique is not good enough in our setting and we use a different approach (which was also used in [56] with slightly different parameters). In Appendix A we explain the approach in detail, compare it to the approach of [6] and give full proofs of the lemmas above.

2.6 A Seeded Bit-Fixing Source Extractor with a Short Seed

In this section we give a construction of a seeded bit-fixing source extractor that uses seed length $O(\log k)$ to extract $k^{\Omega(1)}$ bits as long as k is not too small. This seeded extractor is used as a component in our construction of deterministic extractors for bit-fixing sources.

Theorem 2.6. *Fix any constant* $0 < \alpha < 1$. *There exist constants* $c > 0$ *and* $0 < b < 1$ *(both depending on* α*) such that for any* $n \geq 16$ *and* $k \geq \log^c n$*, there exists an explicit seeded* (k, ϵ)*-bit-fixing source extractor* $E : \{0,1\}^n \times \{0,1\}^d \to \{0,1\}^m$ *with* $d = \alpha \cdot \log k$*,* $m = \Omega(k^b)$ *and* $\epsilon = O(k^{-b})$*.*

Proof. Let X be an (n, k)-bit-fixing source. Let $x = x_1, \ldots, x_n$ be a string sampled by X. The extractor E works as follows: We use the extractor seed y to construct a partition of the bits of x into m sets. Then we output the xor of the bits in each set. With high probability, each set will contain a good bit and therefore, with high probability, the output will be uniformly distributed.

More formally, let b and c be the constants from Lemma 2.18 when using the lemma with the parameter α.

E(x,y)

- We use the seed y to obtain a partition of $[n]$ into $m = \Omega(k^b)$ sets $T_1, \ldots, T_m$ using Lemma 2.18 with the parameter α.
- For $1 \leq i \leq m$, compute $z_i = \oplus_{j \in T_i} x_j$.
- Output $z = z_1, \ldots, z_m$.

We give a detailed correctness proof although it is very straightforward: Let $S \subseteq [n]$ be the set of good indices and let Z be the distribution of the output string z. We need to prove that Z is close to uniform. Let A be the event $\{\forall i \ \ T_i \cap S \neq \emptyset\}$. That is, A is the "good" event in which all sets contain a random bit (and therefore in this case the output is uniform). Let A^c be the complement event, i.e., A^c is the event $\{\exists i \ \ T_i \cap S = \emptyset\}$. We decompose Z according to A and A^c:

$$Z = \Pr(A^c) \cdot (Z|A^c) + \Pr(A) \cdot (Z|A)$$

$(Z|A)$ is uniformly distributed. From Lemma 2.18, when $k \geq \log^c n$, $\Pr(A) \geq 1 - O(k^{-b})$. Therefore, by Lemma 2.8

$$Z \overset{O(k^{-b})}{\sim} U_m.$$

$\square$

2.7 Deterministic Extractors for Bit-Fixing Sources

In this section, we compose the ingredients from previous sections to prove Theorems 2.1 and 2.2. Namely, given choices for a deterministic bit-fixing source extractor, a sampler and a seeded bit-fixing source extractor, we use Theorems 2.3 and 2.4 to get a new deterministic bit-fixing source extractor. This works as follows: We "plug in" a deterministic extractor that extracts a little randomness and a sampler into Theorem 2.4 to get a seed obtainer. We then "plug in" this seed obtainer and a seeded extractor into Theorem 2.3 to get a new deterministic extractor which extracts almost all of the randomness. It is convenient to express this composition as follows:

Theorem 2.7. *Assume we have the following ingredients:*

- *An $(n, k, k_{min}, k_{max}, \delta)$-sampler Samp* $: \{0,1\}^t \rightarrow P([n])$,
- *a deterministic (k_{min}, ϵ^*)-bit-fixing source extractor* $E^* : \{0,1\}^n \rightarrow \{0,1\}^{m'}$ *and*
- *a seeded $(k-k_{max}, \epsilon_1)$-bit-fixing source extractor* $E_1 : \{0,1\}^n \times \{0,1\}^d \rightarrow \{0,1\}^m$

where $m' \geq d + t$. Then we construct a deterministic (k, ϵ)-bit-fixing source extractor $E : \{0,1\}^n \rightarrow \{0,1\}^m$ where $\epsilon = \epsilon_1 + 3 \cdot \max(\epsilon^* + \delta, \epsilon^* \cdot 2^{t+1})$.

Proof. We use *Samp* and E^* in Theorem 2.4 to get a $(k, k - k_{max}, \max(\epsilon^* + \delta, \epsilon^* \cdot 2^{t+1}))$-seed obtainer $F : \{0,1\}^n \rightarrow \{0,1\}^n \times \{0,1\}^{m'-t}$. Since $m' - t \geq d$, we can use F and E_1 in Theorem 2.3 to obtain a deterministic (k, ϵ)-bit-fixing source extractor $E : \{0,1\}^n \rightarrow \{0,1\}^m$ where $\epsilon = \epsilon_1 + 3 \cdot \max(\epsilon^* + \delta, \epsilon^* \cdot 2^{t+1})$. □

We also require the following construction of a seeded extractor (which is in particular a seeded bit-fixing source extractor).

Theorem 2.8 ([55]). *For any n, k and $\epsilon > 0$, there exists a (k, ϵ)-extractor Ext* $: \{0,1\}^n \times \{0,1\}^d \rightarrow \{0,1\}^m$ *where $m = k$ and $d = O(\log^2 n \cdot \log(1/\epsilon) \cdot \log k)$*

2.7.1 An Extractor for Large k (Proof of Theorem 2.1)

To prove Theorem 2.1, we first state results about the required ingredients and then use the ingredients in Theorem 2.7.

We use the deterministic bit-fixing source extractor of Kamp and Zuckerman [37]. Loosely speaking, the following theorem states that when $k >> \sqrt{n}$, we can deterministically extract a polynomial fraction of the randomness with an exponentially small error.

Theorem 2.9 ([37]). *Fix any integers n, k such that $k = b \cdot n^{1/2+\gamma}$ for some $b > 0$ and $0 < \gamma \leq 1/2$. There exists a constant $c > 0$ (*not *depending on any of the parameters) such that there exists an explicit deterministic (k, ϵ^*)-bit-fixing source extractor $E^* : \{0,1\}^n \rightarrow \{0,1\}^m$ where $m = cb^2 \cdot n^{2\gamma}$ and $\epsilon^* = 2^{-m}$.*

Using the theorem above we can obtain a seed of length $O(n^{2\gamma})$. This means that we can afford these many bits for our sampler and seeded bit-fixing source extractor. We use the sampler based on ℓ-wise independence from Lemma 2.16. We use the seeded extractor of [55] (Theorem 2.8), which we now restate in the following form:

Corollary 2.10. *Fix any constants $0 < \gamma \leq 1/2$ and $\alpha > 0$. There exists a constant n' depending on γ such that for any integers n, k satisfying $n > n'$ and $k \leq n$ there exists a (k, ϵ_1)-extractor $E_1 : \{0,1\}^n \times \{0,1\}^d \rightarrow \{0,1\}^m$ where $m = k$, $d = \alpha \cdot n^{2\gamma}$ and $\epsilon_1 = 2^{-\Omega(\alpha \cdot n^\gamma)}$.*

Proof. We use the extractor of Theorem 2.8. We need $d = c_1 \cdot (\log^3 n \cdot \log(1/\epsilon_1))$ random bits for some constant $c_1 > 0$. We want to use at most $\alpha \cdot n^{2\gamma}$ random bits. We get the inequality $\alpha \cdot n^{2\gamma} \geq c_1 \cdot \log^3 n \cdot \log(1/\epsilon_1)$. Equivalently, $\epsilon_1 \geq 2^{-\frac{\alpha \cdot n^{2\gamma}}{c_1 \cdot \log^3 n}}$. So for a large enough n (depending on γ), we can take $\epsilon_1 = 2^{-\frac{\alpha \cdot n^\gamma}{c_1}} = 2^{-\Omega(\alpha \cdot n^\gamma)}$. □

We now compose the ingredients from Theorem 2.9, Lemma 2.16 and Corollary 2.10 to prove Theorem 2.1. The composition is a bit cumbersome in terms of the different parameters. The main issue is that when $k = n^{1/2+\gamma}$, the deterministic extractor of Kamp and Zuckerman extracts $\Omega(n^{2\gamma})$ random bits; and this is enough to use as a seed for a sampler and seeded extractor (that extracts all the randomness) with error $2^{-\Omega(n^\gamma)}$.

Proof. (of Theorem 2.1) Let c be the constant in Theorem 2.9. We use Theorem 2.7 with the following ingredients:

- The $(n, k, (n^{1/2+\gamma})/6, n^{1/2+\gamma}, \delta = 2^{-\Omega(n^\gamma)})$-sampler $Samp : \{0,1\}^t \rightarrow P([n])$ from Lemma 2.16 where $t = (c/72)n^{2\gamma}$.
- The deterministic $((n^{1/2+\gamma})/6, \epsilon^* = 2^{-m'})$-bit-fixing source extractor $E^* : \{0,1\}^n \rightarrow \{0,1\}^{m'}$ from Theorem 2.9 where $m' = (c/36)n^{2\gamma}$.
- The $(k - n^{1/2+\gamma}, \epsilon_1 = 2^{-\Omega(n^\gamma)})$-extractor $E_1 : \{0,1\}^n \times \{0,1\}^d \rightarrow \{0,1\}^m$ from Corollary 2.10 with $d \leq (c/72)n^{2\gamma}$ and $m = k - n^{1/2+\gamma}$.

Note that all three objects exist for a large enough n depending only on γ (c is a universal constant). Note that $m' \geq t+d$. Therefore, applying Theorem 2.7, we get a deterministic (k, ϵ)-bit-fixing source extractor $E : \{0,1\}^n \rightarrow \{0,1\}^m$ where $m = k - n^{1/2+\gamma}$ and

$$\epsilon = \epsilon_1 + 3 \cdot \max(\epsilon^* + \delta, \epsilon^* \cdot 2^{t+1})$$

$$= 2^{-\Omega(n^\gamma)} + 3 \cdot \max\left(2^{-(c/36)n^{2\gamma}} + 2^{-\Omega(n^\gamma)}, 2^{-(c/36)n^{2\gamma}} \cdot 2^{(c/72)n^{2\gamma}+1}\right)$$

$$= 2^{-\Omega(n^\gamma)} + 3 \cdot \max\left(2^{-\Omega(n^\gamma)}, 2^{-(c/72)n^{2\gamma}+1}\right) = 2^{-\Omega(n^\gamma)}$$

(for a large enough n depending on γ). □

2.7.2 An Extractor for Small k (Proof of Theorem 2.2)

To prove Theorem 2.2 we need a deterministic bit-fixing source extractor for $k < \sqrt{n}$. We use the extractor of Theorem 2.5. We prove the theorem in two steps. First, we use Theorem 2.7 to convert the initial extractor into a deterministic bit-fixing source extractor that extracts more bits. We then apply Theorem 2.7 again to obtain a deterministic bit-fixing source extractor that extracts almost all bits.

The following lemma implements the first step and shows how to extract a polynomial fraction of the randomness with a polynomially small error whenever $k \geq \log^c n$ for some constant c.

Lemma 2.19. *There exist constants $c, b > 0$ such that for any $k \geq \log^c n$ and large enough n, there exists an explicit deterministic (k, k^{-b})-bit-fixing source extractor $E : \{0,1\}^n \to \{0,1\}^m$ where $m = k^{\Omega(1)}$.*

Proof. Roughly speaking, the main issue is that we can get $\Omega(\log k)$ random bits using the deterministic extractor of Theorem 2.5. We will need $c_1 \cdot \log\log n$ random bits to use the sampler of Lemma 2.17 and the seeded extractor of Theorem 2.6 (for some constant c_1). Thus, when $k \geq \log^c n$ for large enough c, we will have enough bits.

Formally, we use Theorem 2.7 with the following ingredients:

- The $(n, k, k^e/2, 3 \cdot k^e, \delta = k^{-\Omega(1)})$-sampler $Samp : \{0,1\}^t \to P([n])$ from Lemma 2.17 where $t = \log k/32$ and $e > 1/2$ is the constant from that lemma.
- The deterministic $(k^e/2, \epsilon^* = 2^{-\sqrt{k^e/2}})$-bit-fixing source extractor $E^* : \{0,1\}^n \to \{0,1\}^{m'}$ from Theorem 2.5 where $m' = \log(k^e/2)/4$.
- The seeded $(k - 3 \cdot k^e, \epsilon_1 = (k - 3 \cdot k^e)^{-\Omega(1)})$-bit-fixing source extractor $E_1 : \{0,1\}^n \times \{0,1\}^d \to \{0,1\}^m$ from Theorem 2.6 with $d = \log k/32$ and $m = (k - 3 \cdot k^e)^{\Omega(1)}$.

Note that all three objects exist for $k \geq \log^c n$ for some constant c and large enough n. Assume that n is large enough so that $k \geq \log^c n \geq 2$. To use Theorem 2.7 we need to check that $m' \geq t + d$: Indeed, $m' = \log(k^e/2)/4 \geq \log k/16 = t + d$ (where we used $e > 1/2$, as stated in Lemma 2.17). Applying Theorem 2.7, we get a deterministic (k, ϵ)-bit-fixing source extractor $E : \{0,1\}^n \to \{0,1\}^m$. Notice that for large enough n, $\epsilon_1 = k^{-\Omega(1)}$; therefore

$$\epsilon = \epsilon_1 + 3 \cdot \max(\epsilon^* + \delta, \epsilon^* \cdot 2^{t+1})$$

$$= k^{-\Omega(1)} + 3 \cdot \max\left(2^{-\sqrt{k^e/2}} + k^{-\Omega(1)}, 2^{-\sqrt{k^e/2}} \cdot 2^{\log k/32+1}\right) = k^{-\Omega(1)}$$

(for a large enough n). Also, $m = (k - 3 \cdot k^e)^{\Omega(1)} = k^{\Omega(1)}$ (for a large enough n) so we get the required parameters. □

We now compose the ingredients from Lemmas 2.17 and 2.19 and Theorem 2.8 to prove Theorem 2.2. The composition is a bit cumbersome in terms of the different parameters. The main issue is that we can extract $k^{\Omega(1)}$ random bits using the deterministic extractor of Lemma 2.19. We want $\log^5 n$ random bits to use the seeded extractor of Theorem 2.8. Thus, when $k \geq \log^c n$ for large enough c, we will have enough bits.

Proof. (of Theorem 2.2) Let b be the constant in Lemma 2.19. We use Theorem 2.7 with the following ingredients:

- The $(n, k, k^e/2, 3 \cdot k^e, \delta = k^{-\Omega(1)})$-sampler $Samp : \{0,1\}^t \to P([n])$ from Lemma 2.17 where $t = (b/2)\log k$ and $e > 1/2$ is the constant from that lemma.
- The deterministic $(k^e/2, \epsilon^* = (k^e/2)^{-b})$-bit-fixing source extractor $E^* : \{0,1\}^n \to \{0,1\}^{m'}$ from Lemma 2.19 where $m' = (k^e/2)^{\Omega(1)}$.
- The $(k - 3 \cdot k^e, \epsilon_1 = 1/n)$-extractor $E_1 : \{0,1\}^n \times \{0,1\}^d \to \{0,1\}^m$ from Theorem 2.8 with $d \leq \log^5 n$ and $m = (k - 3 \cdot k^e)$.

Note that all three objects exist for $k \geq \log^c n$ for some constant c and for large enough n. To use Theorem 2.7 we need to check that $m' \geq t + d$; note that $m' = k^{\Omega(1)}$. We take c large enough so that for large enough n $m'/2 > \log^5 n$ and $m'/2 > (b/2)/\log k$. So for such n

$$m' \geq \log^5 n + (b/2)\log k \geq d + t.$$

Applying Theorem 2.7, we get a deterministic (k, ϵ)-bit-fixing source extractor $E : \{0,1\}^n \to \{0,1\}^m$, where

$$\epsilon = \epsilon_1 + 3 \cdot \max(\epsilon^* + \delta, \epsilon^* \cdot 2^{t+1})$$

$$= 1/n + 3 \cdot \max\left((k^e/2)^{-b} + k^{-\Omega(1)}, 2 \cdot (k^e/2)^{-b} \cdot k^{b/2}\right) = k^{-\Omega(1)}$$

(for large enough n). Since $m = k - O(k^e)$ where $1/2 < e < 1$ we are done. □

2.8 Discussion and Open Problems

We give explicit constructions of deterministic bit-fixing source extractors that extract almost all the randomness. However, we achieve rather large error $\epsilon = k^{-\Omega(1)}$ in the case where $k < \sqrt{n}$. We now explain why this happens and suggest how to reduce the error. Recall that in this case our final extractor is based on an initial extractor that extracts only $m = O(\log k)$ bits. When transforming the initial extractor into the final extractor we use the output bits of the initial extractor as a seed for an averaging sampler. The error parameter δ of an averaging sampler has to be larger than 2^{-m}, and as this error is "inherited" by the final extractor we can only get error about $1/k$. A natural way to improve our result is to find a better construction for the initial extractor.

Some applications of deterministic bit-fixing source extractors in adaptive settings of exposure-resilient cryptography require extractors with $\epsilon \ll 2^{-m}$. We do not achieve this goal (even in our first construction that has relatively small error (unless we artificially shorten the output)). Suppose one wants to extract $m = k - u$ bits (for some parameter u). It is interesting to investigate how small the error can be as a function of u? We point out that the existential nonexplicit result achieves error $\epsilon \geq 2^{-u}$ and thus cannot achieve $\epsilon < 2^{-m}$ when $m \geq k/2$. We remark that for bit-fixing sources we have examples of settings where the nonexplicit result is not optimal. For example, when $m = 1$ the xor-extractor is errorless (see also [15]). Given the discussion above we find it interesting to achieve $m = \Omega(k)$ with $\epsilon = 2^{-\Omega(k)}$ for every choice of k.

Chapter 3

Deterministic Extractors for Affine Sources over Large Fields

Summary

An (n,k)-*affine source* over a finite field $\mathbb{F}$ is a random variable $X = (X_1, ..., X_n) \in \mathbb{F}^n$, which is uniformly distributed over an (unknown) k-dimensional affine subspace of $\mathbb{F}^n$. We show how to (deterministically) extract practically all the randomness from affine sources, for any field of size larger than n^c (where c is a large enough constant). This chapter is based on [25].

3.1 Introduction

Let $\mathbb{F}$ be a finite field of size q and let n be an integer. The famous Hales-Jewett theorem [35] implies that if n is large enough compared to q then in any two-coloring of the vector space $\mathbb{F}^n$ there exists a monochromatic line[1]. On the other hand, if q is significantly larger than n (say, $q \geq 3n \log_2 n$) then a random two-coloring of the vector space $\mathbb{F}^n$ doesn't have monochromatic lines (with high probability). Assume that q is large enough (say, $q \geq n^{20}$). Can one give an *explicit* two-coloring of $\mathbb{F}^n$ that doesn't have monochromatic lines ? More generally, can one give an explicit coloring $D : \mathbb{F}^n \rightarrow \{0,1\}$ such that every line will have roughly the same number of 0s and 1 ?

The problem of extracting randomness from affine sources is a more general problem. Fix n, k and $\mathbb{F}$. Assume that X is uniformly distributed over an *unknown* k-dimensional affine subspace of $\mathbb{F}^n$. The goal is to give an explicit example for a function $D : \mathbb{F}^n \rightarrow \Omega$ (for some finite set Ω) such that the distribution of $D(X)$ is ϵ-close to uniform. Naturally, we would like Ω to be as large as possible and ϵ to be as small as possible.

[1] A line is a 1-dimensional affine subspace of $\mathbb{F}^n$.

A. Gabizon, *Deterministic Extraction from Weak Random Sources*,
Monographs in Theoretical Computer Science. An EATCS Series,
DOI 10.1007/978-3-642-14903-0_3,

3.1.1 Affine Source Extractors

Denote by $\mathbb{F}_q$ the finite field with q elements. Denote by $\mathbb{F}_q^n$ the n-dimensional vector space over $\mathbb{F}_q$.

Definition 3.1 (affine source). *A distribution X over $\mathbb{F}_q^n$ is an $(n,k)_q$-affine source if it is uniformly distributed over an affine subspace of dimension k. That is, X is sampled by choosing $t_1, \ldots, t_k$ uniformly and independently in $\mathbb{F}_q$ and calculating*

$$\sum_{j=1}^{k} t_j \cdot a^{(j)} + b$$

for some $a^{(1)}, \ldots, a^{(k)}, b \in \mathbb{F}_q^n$ such that $a^{(1)}, \ldots, a^{(k)}$ are linearly independent.

For a finite set Ω, we denote by U_Ω the uniform distribution on Ω. We recall that two distributions P and Q over Ω are ϵ-close (denoted by $P \overset{\epsilon}{\sim} Q$) if for every event $A \subseteq \Omega$, $|\Pr_P(A) - \Pr_Q(A)| \leq \epsilon$.

Definition 3.2 (deterministic affine source extractor). *A function $D : \mathbb{F}_q^n \to \Omega$ is a deterministic (k, ϵ)-affine source extractor if for every $(n,k)_q$-affine source X the distribution $D(X)$ is ϵ-close to uniform. That is,*[2]

$$D(X) \overset{\epsilon}{\sim} U_\Omega.$$

3.1.2 Our Results

We construct deterministic extractors for affine sources over large fields. Specifically, we work with a field size that is polynomially large in n. We give constructions that extract practically all the randomness in all cases. We have two main constructions. The first is designed for $k \geq 2$ and the second for $k = 1$.

Our first construction gives a deterministic affine source extractor that extracts $k-1$ random elements[3] in $\mathbb{F}_q$ from any $(n,k)_q$-affine source, provided q is a large enough polynomial in n. Note that we didn't make any attempt to optimize the constants 20 and 21 in the following theorem (as they depend on each other).

Theorem 3.1. *There exists a constant q_0 such that for any field $\mathbb{F}_q$ and integers n, k with $q > \max[q_0, n^{20}]$, there is an explicit deterministic (k, ρ)-affine source extractor $D : \mathbb{F}_q^n \to \mathbb{F}_q^{k-1}$ with $\rho \leq q^{-1/21}$.*

---

[2] Our extractors will sometimes output bits and sometimes output field elements. Therefore, the definition here uses a general output domain.

[3] Actually, we can construct a deterministic (k, ϵ)-affine source extractor that outputs $k-1$ random elements in $\mathbb{F}_q$ *and* $\lfloor (1-\delta) \cdot \log q \rfloor$ random bits for any constant $0 < \delta < 1$.

Our second result is for $k = 1$. It gives a deterministic affine source extractor that extracts all the randomness except for an entropy loss of $2\log_2(n/\epsilon) + o(\log_2 q)$ bits.

Theorem 3.2. *For any field $\mathbb{F}_q$, integer n and $\epsilon > 0$, there is an explicit deterministic $(1,\epsilon)$-affine source extractor $D : \mathbb{F}_q^n \to \{0,1\}^d$ with $d = \lfloor \log_2 q - 2\log_2(n/\epsilon) - 2\log_2\log_2 q - 4 \rfloor$.*

We note the following possible instantiations of the theorem.

- Assuming $q > n^c$, we can extract a $(1-\delta)$ fraction of the source randomness, where $\delta > 0$ is an arbitrarily small constant, and c is a constant[4] depending on δ.
- Using any $q \geq n^2 \cdot \log_2^3 n$ and $\epsilon = 1/4$ with a one-bit output, we get an explicit two-coloring of $\mathbb{F}_q$ such that no line is monochromatic.

The main drawback of Theorem 3.1 is the large error. The error that we achieve is polynomially small in q. However, the error ρ does not decrease as k increases. (We might have hoped to have error exponentially small in k.) This is because, as will be explained in Section 3.2, the first stage of our construction extracts randomness from an $(n,1)_q$-affine source. The error of the entire construction is bounded from below by the error of this stage.

3.1.3 Previous Work

Previous works studied the problem over the field $\mathbb{F}_2$ (i.e., $GF[2]$). In [4], Barak, Kindler, Shaltiel, Sudakov and Wigderson show how to extract one non-constant bit for k slightly sub-linear in n. In other words, their result gives a two-coloring of $\mathbb{F}_2^n$ in which no affine subspace of linear dimension (or slightly sub-linear dimension) is monochromatic. More recently, Bourgain[10] showed how to extract $\Omega(k)$ bits that are exponentially close to uniform when k is linear in n.

3.2 Overview of Techniques

The basic scheme of our construction is as follows: We construct a deterministic affine source extractor that extracts a few bits. We then use these bits to run a "seeded extractor" that extracts almost all the randomness from the source. (Usually, seeded extractors require a seed that is independent of the source. We will construct a "special kind" of seeded extractor that can work well even with a seed that is correlated with the affine source). The proof that this composition of extractors works is based on the recycling paradigm described in the introductory chapter. We now elaborate on the components in this scheme.

[4]See Lemma 3.10 for an exact formulation of such an instantiation.

3.2.1 Extracting Many Bits from Lines

As described above, the first step of our construction is to extract a few bits deterministically. We do this by showing a method to extract any constant fraction of the randomness from an $(n, 1)_q$-affine source, assuming $q > n^c$ for large enough c. We first describe how to extract one bit when q is slightly more than quadratic in n.

Extracting a single bit: We want to extract one random bit from an $(n, 1)_q$-affine source, assuming $q = n^{2+\gamma}$ for some $\gamma > 0$. Consider first the easier task of outputting a non-constant bit or even a non-constant value over a larger domain, say $\mathbb{F}_q$. This can be achieved by the following method: Given input $x = (x_1, \ldots, x_n) = (a_1 \cdot t + b_1, \ldots, a_n \cdot t + b_n) \in \mathbb{F}_q^n$ (where $a_i, b_i \in \mathbb{F}_q$ are constant and t is chosen uniformly at random in $\mathbb{F}_q$), we compute the expression $\sum_{i=1}^n x_i^i = \sum_{i=1}^n (a_i \cdot t + b_i)^i$. We know that $a_i \neq 0$ for some i. Assume for simplicity that $a_n \neq 0$. The nth summand is a polynomial of degree n in the variable t. Since the other summands do not contain nth powers, the entire expression is a non-constant polynomial in t (the large field size comes in here). Since t is chosen uniformly in $\mathbb{F}_q$, our output will be non-constant. Actually, by computing this expression we have "converted" our distribution into a "low-degree distribution" of the form $f(U_{\mathbb{F}_q})$, that is, a distribution sampled by choosing t uniformly in $\mathbb{F}_q$ and computing $f(t)$ for some low-degree polynomial f (low-degree in relation to the field size). Noticing this, the way to a random bit becomes easy using well-known theorems of Weil [72] about character sums.[5] The *characters* of a finite field $\mathbb{F}_q$ are functions from $\mathbb{F}_q$ to the complex numbers that preserve the field addition or multiplication (see subsection 3.3.2 for definitions). Weil's theorems show that field characters of order 2 are actually "deterministic extractors" for such "low-degree distributions" (unless the polynomial is of a certain restricted form). Thus, our extractor works by "converting" the source distribution into a "low-degree distribution"[6] $f(U_{\mathbb{F}_q})$, and then applying a character of order 2.

Extracting many bits: As explained in subsection 3.3.2, we will need to work a bit differently for fields of even and odd characteristic. For simplicity, let us consider now the case of an even-sized field. As described in subsection 3.3.2, when q is even, we use Weil's theorems to show that the trace function $Tr : \mathbb{F}_q \to \mathbb{F}_2$ (defined in subsection 3.3.2) outputs an almost unbiased bit when given a sample from a "low-degree distribution" $f(U_{\mathbb{F}_q})$, where f is a polynomial of odd degree. Furthermore, Tr is an additive function; that is, $Tr(a + b) = Tr(a) + Tr(b)$. Our extractor works as follows: In a way similar

[5] These theorems have already been very fruitful in computer science, e.g., in explicit constructions of ϵ-biased spaces [2], tournaments [33, 1] and pseudorandom graphs [45].

[6] We use a slightly different expression than the one given here to ensure that f will not be of a certain restricted form on which Weil's theorems don't apply.

to the one bit case, we use our source to produce samples from several "low-degree distributions" of the form $U(f_j')$ where the f_j's have odd degree. We then apply Tr on each sample. This gives us several bits that are each individually close to uniform. We want to ensure that their joint distribution is also close to uniform. For this purpose, we make sure the f_j's have the property that the sum of any subset of them is also a polynomial of odd degree. We use this property together with the additivity of Tr to show that the parity of any subset of the output bits is close to uniform. We then use the Vazirani Xor Lemma (see, for example, [22]) to conclude that the output distribution is close to uniform. The case of an odd-sized field is similar but requires a bit more work.

3.2.2 Linear Seeded Affine Source Extractors

Our goal is to construct deterministic affine source extractors. As a component in our construction, we use linear *seeded* extractors for affine sources, i.e., seeded extractors that work only on affine sources (and not on general high min-entropy sources). Furthermore, the extractors are linear, meaning that for any fixed seed, the extractor is a linear function of the source.

Definition 3.3 (linear seeded affine source extractor). *A function $E : \mathbb{F}_q^n \times \{0,1\}^d \to \mathbb{F}_q^m$ is a linear seeded (k,ϵ)-affine source extractor if*

1. *For every $(n,k)_q$-affine source X, the distribution $E(X, U_d)$ is ϵ-close to uniform. That is,*
$$E(X, U_d) \overset{\epsilon}{\sim} U_{\mathbb{F}_q^m}.$$
2. *For a fixed seed, E is a linear function. That is, for any $a^{(1)}, a^{(2)} \in \mathbb{F}_q^n, t_1, t_2 \in \mathbb{F}_q$ and $y \in \{0,1\}^d$, we have*
$$E(t_1 \cdot a^{(1)} + t_2 \cdot a^{(2)}, y) = t_1 \cdot E(a^{(1)}, y) + t_2 \cdot E(a^{(2)}, y).$$

We now sketch our construction of linear seeded affine source extractors (see Section 3.6 for full details). Fix any affine subspace $A \subseteq \mathbb{F}_q^n$ of dimension k. It is not hard to show that a random linear mapping $T : \mathbb{F}_q^n \to \mathbb{F}_q^k$, or equivalently, a random $k \times n$ matrix over $\mathbb{F}_q$, will map A (uniformly) onto $\mathbb{F}_q^k$ with probability at least $1 - \frac{1}{q-1}$. Our construction of linear seeded affine source extractors can be viewed as a derandomization of this property. Assuming $q > n^3$, we construct a set of *less than* q matrices with a similar property. That is, for any affine subspace $A \subseteq \mathbb{F}_q^n$ of dimension k, most of the matrices in this set will map A onto $\mathbb{F}_q^k$. The construction is very simple: Pick any subset $U \subseteq \mathbb{F}_q$ with $|U| > n^3$. The set of matrices will be the "power matrices" of the elements of U. That is, for each $u \in U$ we will have a $k \times n$ matrix T_u where $(T_u)_{j,i} = u^{ji}$ (where ji is the product of j and i as integers).

For general high min-entropy sources, it is known that encoding the source string with an error correcting code and outputting random locations of the encoding make a good extractor. Some extractor constructions for general high min-entropy sources, specifically the breakthrough construction of Trevisan [66], and its improvement by Raz, Reingold and Vadhan [55] and also the very elegant constructions of Ta-Shma, Zuckerman and Safra [65] and Shaltiel and Umans[64], can be viewed as using the random seed to select locations from an encoding of the source in a derandomized way. From this angle, our construction may be viewed as selecting locations from the Reed-Solomon encoding[7] of the (affine) source in a derandomized way. Specifically, we choose the first location u randomly from a large enough subset $U \subseteq \mathbb{F}_q$. The other locations are simply the powers of u, i.e., $u^2, u^3, \ldots, u^k$.

Remark 3.1. *We note that some extractor constructions for general high min-entropy sources, for example, the constructions of [55, 64, 65, 66] discussed above, are already linear seeded affine source extractors. They are designed to work over the binary field but seem to be easily adaptable to large fields. Why not use one of these constructions? This is a possibility. However, our construction is considerably simpler and achieves better parameters for the case of affine sources. In particular, using one of the above-mentioned constructions would not have enabled us to extract almost all the randomness (as we will need an affine source extractor that can do so with a seed of length $O(\log n)$).*

3.2.3 Using the Correlated Randomness as a Seed

As stated earlier, we wish to use the few bits extracted by the deterministic affine source extractor D (described in subsection 3.2.1) as a seed for the linear seeded affine source extractor E described in subsection 3.2.2. In principle, this is problematic as a seeded extractor is only guaranteed to work when its seed is independent of the source. We want to use a seed that is a *function* of the source. However, using the recycling paradigm, we show that when the seeded extractor is linear this does work. Let us sketch the argument: Given a fixed seed u, E is a linear mapping. Therefore, if X is an affine source, then given a possible output value a, the distribution X conditioned on $E(x, u) = a$ is *also* an affine source (as we have just added another linear constraint on the support of X). Hence, the distribution $D(X)$, even when conditioned on $E(x, u) = a$, is still close to uniform. Using simple manipulations of probability distributions, this can be used to show that the distribution $E(X, D(X))$ is close to the distribution $E(X, U_d)$ (and therefore close to uniform). See also the description of the recycling paradigm in the first chapter.

[7] The Reed-Solomon encoding of $x = (x_1, \ldots, x_n) \in \mathbb{F}_q^n$ at location $u \in \mathbb{F}_q$ is defined as $\sum_{i=1}^{n} x_i \cdot u^i$.

3.3 Preliminaries

Notation: We use $[n]$ to denote the set $\{1,\ldots,n\}$. Let Ω, Π be some finite sets. For $x \in \Omega^n$ and $i \in [n]$, we denote by x_i the ith coordinate of x. Similarly, for a function $D : \Pi \to \Omega^n$ and $i \in [n]$, we denote by D_i the function D restricted to the ith output coordinate. Logarithms will always be taken to base 2. We denote by $\mathbb{F}_q$ the finite field of q elements. We denote by $\overline{\mathbb{F}}_q$ the algebraic closure of $\mathbb{F}_q$ and by $\mathbb{F}_q[t]$ the ring of formal polynomials over $\mathbb{F}_q$. We denote by $\mathbb{F}_q^n$ the vector space of dimension n over $\mathbb{F}_q$. Given a $k \times n$ matrix T over $\mathbb{F}_q$, we also view T as a mapping from $\mathbb{F}_q^n$ to $\mathbb{F}_q^k$ and denote $T(x) \triangleq T \cdot x$ for $x \in \mathbb{F}_q^n$.

3.3.1 Probability Distributions

Notation for probability distributions: Let Ω be some finite set. Let P be a distribution on Ω. For $B \subseteq \Omega$, we denote the probability of B according to P by $\Pr_P(B)$ or $\Pr(P \subseteq B)$; When $B \in \Omega$, we will also use the notation $\Pr(P = B)$. Given a function $A : \Omega \to U$, we denote by $A(P)$ or by $[A(t)]_{t \leftarrow P}$ the distribution induced on U when sampling t by P and calculating $A(t)$. We will use the same notation for expressions not explicitly named as functions. For example, for a distribution P on $\mathbb{F}_q$ we will denote by $P+1$ or by $[t+1]_{t \leftarrow P}$ the distribution induced on $\mathbb{F}_q$ by sampling t by P and adding 1. When we write $t_1, \ldots, t_k \leftarrow P$, we mean that $t_1, \ldots, t_k$ are chosen *independently* according to P. We denote by U_Ω the uniform distribution on Ω. For an integer n, we denote by U_n the uniform distribution on $\{0,1\}^n$. We abuse notation and denote by U_q the uniform distribution on $\mathbb{F}_q$. In any expression involving U_Ω or U_n and other distributions, the instance of U_n or U_Ω is independent of the other distributions. For a distribution P on Ω^d and $j \in [d]$, we denote by P_j the restriction of P to the jth coordinate. We denote by *Supp*(P) the support of P. The *statistical distance* between two distributions P and Q on Ω, denoted by $|P - Q|$, is defined as

$$|P - Q| \triangleq \max_{S \subseteq \Omega} \left| \Pr_P(S) - \Pr_Q(S) \right| = \frac{1}{2} \sum_{w \in \Omega} \left| \Pr_P(w) - \Pr_Q(w) \right| .$$

We say that P is ϵ-*close* to Q, denoted by $P \overset{\epsilon}{\sim} Q$, if $|P - Q| \leq \epsilon$. We denote the fact that P and Q are identically distributed by $P \sim Q$.

We define conditional distributions.

Definition 3.4 (conditional distributions). *Let P be a distribution on Ω. Let $C \subseteq \Omega$ be an event such that $\Pr_P(C) > 0$. We define the distribution $(P|C)$ by*

$$\Pr_{(P|C)}(B) = \frac{\Pr_P(B \cap C)}{\Pr_P(C)}$$

for any $B \subseteq \Omega$. Given a function $A : \Omega \to U$, we denote by $(A(P)|C)$ the distribution $A((P|C))$.

We will need the notion of a convex combination of distributions.

Definition 3.5 (convex combination of distributions). *Given distributions $P_1, \dots, P_t$ on a set Ω and coefficients $\mu_1, \dots, \mu_t \geq 0$ such that $\sum_{i=1}^t \mu_i = 1$, we define the distribution $P \triangleq \sum_{i=1}^t \mu_i \cdot P_i$ by*

$$\Pr_P(B) = \sum_{i=1}^t \mu_i \cdot \Pr_{P_i}(B)$$

for any $B \subseteq \Omega$.

We will need a few technical lemmas on probability distributions.

The following lemma shows that convex combinations of similar distributions with similar coefficients are statistically close.

Lemma 3.2. *Let t be any integer. Let $P_1, \dots, P_t$ and $Q_1, \dots, Q_t$ be sequences of distributions on a set Ω such that for every $i \in [t]$, $P_i \overset{\epsilon}{\sim} Q_i$. Let μ and ν be distributions on $[t]$ with $|\mu - \nu| \leq \delta$. Let $P \triangleq \sum_{i=1}^t \Pr(\mu = i) \cdot P_i$, $Q \triangleq \sum_{i=1}^t \Pr(\nu = i) \cdot Q_i$. Then $P \overset{2\cdot\delta+\epsilon}{\sim} Q$.*

Proof. Denote $\mu_i = \Pr(\mu = i)$ and $\nu_i = \Pr(\nu = i)$. Given $B \subseteq \Omega$, we have

$$\left|\Pr_P(B) - \Pr_Q(B)\right| = \left|\sum_{i=1}^t \mu_i \cdot \Pr_{P_i}(B) - \sum_{i=1}^t \nu_i \cdot \Pr_{Q_i}(B)\right|$$

$$\leq \sum_{i=1}^t \left|\mu_i \cdot \Pr_{P_i}(B) - \nu_i \cdot \Pr_{Q_i}(B)\right| \leq \sum_{i=1}^t \left|\mu_i \cdot \Pr_{P_i}(B) - \nu_i \cdot \Pr_{P_i}(B) + \nu_i \cdot \Pr_{P_i}(B) - \nu_i \cdot \Pr_{Q_i}(B)\right|$$

$$\leq \sum_{i=1}^t |\mu_i - \nu_i| + \nu_i \left|\Pr_{P_i}(B) - \Pr_{Q_i}(B)\right| \leq 2\cdot\delta + \sum_{i=1}^t \nu_i \left|\Pr_{P_i}(B) - \Pr_{Q_i}(B)\right| \leq 2\cdot\delta + \epsilon.$$

□

Lemma 3.3. *Let $P_1, \dots, P_t$ be a sequence of distributions on a set Ω. Let μ be a distribution on $[t]$. Let $P \triangleq \sum_{i=1}^t \Pr(\mu = i) \cdot P_i$. Assume that the probability given by μ to the non-uniform P_is is at most ϵ, i.e., $\Pr_{i \leftarrow \mu}(P_i \nsim U_\Omega) \leq \epsilon$. Then*

$$P \overset{\epsilon}{\sim} U_\Omega.$$

Proof. By the assumption of the lemma, $P = (1 - \delta) \cdot U_\Omega + \delta \cdot V$ for some $\delta \leq \epsilon$ and distribution V on Ω. Let $B \subseteq \Omega$ be some event.

$$\left|\Pr_P(B) - \Pr_{U_\Omega}(B)\right| = \left|\delta \cdot \Pr_V(B) + (1-\delta) \cdot \Pr_{U_\Omega}(B) - \Pr_{U_\Omega}(B)\right| \leq \delta \cdot \left|\Pr_V(B) - \Pr_{U_\Omega}(B)\right| \leq \delta \leq \epsilon.$$

□

3.3.2 Characters of Finite Fields

Given an abelian group G, a *character* on G is a map from G to complex roots of unity that preserves the group action. The characters of a finite field are the characters of the additive and multiplicative[8] groups of the field.

Definition 3.6 (additive character). *A function $\psi : \mathbb{F}_q \to \mathbb{C}$ is an additive character of $\mathbb{F}_q$ if $\psi(0) = 1$ and*

$$\psi(a+b) = \psi(a)\psi(b)$$

for every $a, b \in \mathbb{F}_q$. The order of ψ is the smallest integer d such that $(\psi(a))^d = 1$ for every $a \in \mathbb{F}_q$.

Definition 3.7 (multiplicative character). *A function $\chi : \mathbb{F}_q \to \mathbb{C}$ is a multiplicative character of $\mathbb{F}_q$ if $\chi(1) = 1$, $\chi(0) = 0$ and*

$$\chi(ab) = \chi(a)\chi(b)$$

for every $a, b \in \mathbb{F}_q$. The order of χ is the smallest integer d such that $(\chi(a))^d = 1$ for every $a \in \mathbb{F}_q^*$.

We will concentrate on characters of order 2. Even-sized fields have additive characters of order 2 and odd-sized fields have a multiplicative character of order 2. We define a character of order 2 for each case and also a "boolean version" of the character (i.e., a function with range $\{0, 1\}$) that we will use in our extractor construction.

Definition 3.8 (additive character of order 2). *Let $q = 2^l$ for some integer l. The function $Tr : \mathbb{F}_q \to \mathbb{F}_2$ is defined to be the trace of $\mathbb{F}_q$ over $\mathbb{F}_2$. That is,*[9]

$$Tr(a) = a + a^2 + a^{2^2} + \ldots + a^{2^{l-1}}.$$

We define the additive character $\psi_1 : \mathbb{F}_q \to \{1, -1\}$ by[10] *$\psi_1(a) = -1^{Tr(a)}$.*

Definition 3.9 (Multiplicative character of order 2). *Let $q = p^l$ for some integer l and odd prime p. We define the multiplicative character $\chi_1 : \mathbb{F}_q \to \{-1, 0, 1\}$ to be 1 for a nonzero quadratic residue, -1 for a quadratic non-residue, and 0 on 0. More concisely,*

$$\chi_1(a) = a^{\frac{q-1}{2}}.$$

We define the function $QR : \mathbb{F}_q \to \{0, 1\}$ by $QR(a) = 1$ if $\chi_1(a) = -1$, and $QR(a) = 0$ otherwise. That is, $QR(a) = 1$ for quadratic non-residues and 0 otherwise.

---

[8] A character χ of $\mathbb{F}_q^*$ is extended to 0 by $\chi(0) = 0$.
[9] It is known that $Tr(a) \in \mathbb{F}_2$ for every $a \in \mathbb{F}_q$.
[10] We interpret the field elements 0 and 1 as the corresponding integers.

It is obvious that χ_1 and ψ_1 have order at most 2. It can be shown that their order is exactly 2.

Very useful theorems of Weil [72] state that for any low-degree polynomial f that is not of a certain restricted form, the values of a field character "cancel out" over the range of f (when viewed as a multi-set). We state two special cases of these theorems. The theorems can be found in [59]. The first theorem deals with additive characters.

Theorem 3.3. *[59][Theorem 2E, page 44] Let ψ be a nontrivial additive character of $\mathbb{F}_q$ (that is, not identically 1). Let $f(t)$ be a polynomial in $\mathbb{F}_q[t]$ of degree m. Suppose that $gcd(m,q)=1$. Then*

$$\left|\sum_{t\in\mathbb{F}_q}\psi(f(t))\right| \le mq^{1/2}.$$

The second theorem deals with multiplicative characters.

Theorem 3.4. *[59][Theorem 2C′, page 43] Let χ be a multiplicative character of $\mathbb{F}_q$ of order $d>1$. Let $f(t)$ be a polynomial in $\mathbb{F}_q[t]$ of degree m. Suppose that $f(t)$ is not of the form $c\cdot g(t)^d$ for any $c\in\mathbb{F}_q$ and $g(t)\in\mathbb{F}_q[t]$. Then*

$$\left|\sum_{t\in\mathbb{F}_q}\chi(f(t))\right| \le mq^{1/2}.$$

For the case of a field character of order 2, Weil's theorems actually show that the character is a "deterministic extractor"[11] for distributions of the form $f(U_q)$ for almost any low-degree polynomial f. We formalize this in the following corollaries of Theorems 3.3 and 3.4 stated for the boolean versions of the characters ψ_1 and χ_1.

Corollary 3.5. *Let q be a power of 2. Let $f\in\mathbb{F}_q[t]$ be a polynomial of odd degree m. Then*

$$Tr(f(U_q)) \overset{\frac{m}{2\sqrt{q}}}{\sim} U_1.$$

Proof.

$$\left|\sum_{t\in\mathbb{F}_q}\psi_1(f(t))\right| = \left|\sum_{t\in\mathbb{F}_q,\psi_1(f(t))=1} 1 - \sum_{t\in\mathbb{F}_q,\psi_1(f(t))=-1} 1\right|$$

$$= q\cdot\left|\Pr_{t\leftarrow U_q}(\psi_1(f(t))=1) - \Pr_{t\leftarrow U_q}(\psi_1(f(t))=-1)\right| = q\cdot\left|\Pr_{t\leftarrow U_q}(Tr(f(t))=0) - \Pr_{t\leftarrow U_q}(Tr(f(t))=1)\right|$$

$$= q\cdot\left|2\cdot\Pr_{t\leftarrow U_q}(Tr(f(t))=0)-1\right| = 2q\cdot\left|\Pr_{t\leftarrow U_q}(Tr(f(t))=0)-1/2\right| = 2q\cdot|Tr(f(U_q))-U_1|.$$

[11] Characters of higher order are also extractors, but with larger error.

Since $gcd(m,q)=1$, using Theorem 3.3 we have

$$|Tr(f(U_q))-U_1| = \frac{1}{2q}\cdot\left|\sum_{t\in\mathbb{F}_q}\psi_1(f(t))\right| \leq \frac{1}{2q}\cdot mq^{1/2} = \frac{m}{2\sqrt{q}}.$$

□

The proof of the analogous claim for χ_1 is a bit more cumbersome as we have to deal with the artificial extension of χ_1 to $\mathbb{F}_q$ by $\chi_1(0)=0$. We will use the following definition.

Definition 3.10 (square multiple). *We say that a polynomial $f(t)$ in $\mathbb{F}_q[t]$ is a square multiple in $\mathbb{F}_q[t]$ if $f(t)=c\cdot g(t)^2$ for some $c\in\mathbb{F}_q$ and $g(t)\in\mathbb{F}_q[t]$.*

Corollary 3.6. *Let $q=p^l$ for some integer l and odd prime p. Let $f(t)\in\mathbb{F}_q[t]$ be a polynomial of degree m that is not a square multiple in $\mathbb{F}_q[t]$. Then*

$$QR(f(U_q)) \overset{\frac{m}{\sqrt{q}}}{\sim} U_1.$$

Proof. We have

$$\sum_{t\in\mathbb{F}_q}\chi_1(f(t)) = \left[\sum_{t\in\mathbb{F}_q,\chi_1(f(t))=1} 1 - \sum_{t\in\mathbb{F}_q,\chi_1(f(t))=-1} 1\right]$$

$$= q\cdot\left[\Pr_{t\leftarrow U_q}(\chi_1(f(t))=1) - \Pr_{t\leftarrow U_q}(\chi_1(f(t))=-1)\right]$$

$$= q\cdot\left[\Pr_{t\leftarrow U_q}(QR(f(t))=0) - \Pr_{t\leftarrow U_q}(f(t)=0) - \Pr_{t\leftarrow U_q}(QR(f(t))=1)\right]$$

$$= q\cdot\left[2\cdot\Pr_{t\leftarrow U_q}(QR(f(t))=0)-1\right] - q\cdot\Pr_{t\leftarrow U_q}(f(t)=0)$$

$$=2q\cdot\left[\Pr_{t\leftarrow U_q}(QR(f(t))=0)-1/2\right]-q\cdot\Pr_{t\leftarrow U_q}(f(t)=0)=2q|QR(f(U_q))-U_1|-q\cdot\Pr_{t\leftarrow U_q}(f(t)=0)$$

where in the last equality we assumed without loss of generality that

$$\Pr_{t\leftarrow U_q}(QR(f(t))=0)\geq 1/2.$$

Since χ_1 is of order 2 and $f(t)$ is not of the form $c\cdot g(t)^2$ for any $c\in\mathbb{F}_q$ and $g(t)\in\mathbb{F}_q[t]$, using Theorem 3.4 we have

$$|QR(f(U_q))-U_1| = \frac{1}{2q}\cdot\sum_{t\in\mathbb{F}_q}\chi_1(f(t)) + (1/2)\cdot\Pr_{t\leftarrow U_q}(f(t)=0)$$

$$\leq\frac{1}{2q}\cdot mq^{1/2}+\frac{m}{2q}\leq\frac{m}{2\sqrt{q}}+\frac{m}{2\sqrt{q}}=\frac{m}{\sqrt{q}}.$$

□

3.4 Extracting One Bit from Lines

In the next section we show how to extract any constant fraction of the randomness from an $(n,1)_q$-affine source, provided q is a large enough polynomial in n. For simplicity of the presentation, we first show how to extract one bit from an $(n,1)_q$-affine source when q is slightly more than quadratic in n.

As explained in Section 3.2, we first "convert" a uniform distribution on a one-dimensional affine subspace into a distribution of the form $f'(U_q)$, where f' is a low-degree polynomial; we then apply a (boolean version of a) field character of order 2. Weil's theorems guarantee that our output will be close to uniform. As explained in subsection 3.3.2, since we want a field character of order 2 we need to use an additive character for even-sized fields and a multiplicative character for odd-sized fields.

The following lemma shows how to extract one bit when the field size is even.

Lemma 3.4. *Let q be a power of 2. Fix any integer $n < \sqrt{q}$. Define the multivariate polynomial $f : \mathbb{F}_q^n \to \mathbb{F}_q$ by $f(x) = \sum_{i=1}^n x_i^{2i-1}$. The function $D_0 : \mathbb{F}_q^n \to \{0,1\}$ defined by $D_0(x) = Tr(f(x))$ is a deterministic $(1,\epsilon)$-affine source extractor, where $\epsilon = n/\sqrt{q}$.*

Proof. Fix an $(n,1)_q$-affine source X. Recall that $X \sim [t \cdot a + b]_{t \leftarrow U_q}$ for some $a, b \in \mathbb{F}_q^n$ such that $a \neq 0$. We have

$$D_0(X) \sim Tr(f(X)) \sim [Tr(f(t \cdot a_1 + b_1, \ldots, t \cdot a_n + b_n))]_{t \leftarrow U_q}$$

$$\sim \left[Tr\left(\sum_{i=1}^{n} (t \cdot a_i + b_i)^{2i-1} \right) \right]_{t \leftarrow U_q}.$$

Denote $f'(t) = \sum_{i=1}^n (t \cdot a_i + b_i)^{2i-1}$. Note that f' is a polynomial of odd degree m, where $m \leq 2n$. Therefore, using Corollary 3.5 we have

$$D_0(X) \sim Tr(f'(U_q)) \overset{\frac{n}{\sqrt{q}}}{\sim} U_1.$$

□

The following lemma shows how to extract one bit when the field size is odd.

Lemma 3.5. *Let $q = p^l$ for some integer l and odd prime p. Fix any integer $n < \sqrt{q}/2$. Define the multivariate polynomial $f : \mathbb{F}_q^n \to \mathbb{F}_q$ by $f(x) = \sum_{i=1}^n x_i^{2i-1}$. The function $D_0 : \mathbb{F}_q^n \to \{0,1\}$ defined by $D_0(x) = QR(f(x))$ is a deterministic $(1,\epsilon)$-affine source extractor, where $\epsilon = 2n/\sqrt{q}$.*

Proof. Fix an $(n,1)_q$-affine source $X \sim [t \cdot a + b]_{t \leftarrow U_q}$. We have

$$D_0(X) \sim QR(f(X)) \sim [QR(f(t \cdot a_1 + b_1, \ldots, t \cdot a_n + b_n))]_{t \leftarrow U_q}$$
$$\sim \left[QR\left(\sum_{i=1}^{n}(t \cdot a_i + b_i)^{2i-1}\right)\right]_{t \leftarrow U_q}.$$

Denote $f'(t) = \sum_{i=1}^{n}(t \cdot a_i + b_i)^{2i-1}$. Note that $f'(t)$ is a polynomial of odd degree m (and therefore not a square multiple in $\mathbb{F}_q[t]$) where $m \le 2n$. Therefore, using Corollary 3.6 we have

$$D_0(X) \sim QR(f'(U_q)) \overset{\frac{2n}{\sqrt{q}}}{\sim} U_1.$$

□

3.5 Extracting Many Bits from Lines

In this section we prove Theorem 3.2. In particular, we show how to extract any constant fraction of the randomness from an $(n,1)_q$-affine source provided q is a large enough polynomial in n. We will prove the correctness of our construction by showing that the parity of any subset of the output bits is almost unbiased. The following "Xor Lemma" due to Vazirani states that this indeed implies that the output is close to uniform. The lemma follows from elementary Fourier analysis. For a proof see [22].

Lemma 3.6. *Let X be a distribution on $\{0,1\}^d$. Assume that for every nonempty subset $S \subseteq [d]$*

$$\bigoplus_{j \in S} X_j \overset{\epsilon}{\sim} U_1$$

(where $\bigoplus$ denotes addition mod 2). Then

$$|X - U_d| \le \epsilon \cdot 2^{d/2}.$$

We first deal with fields of even size. As explained in Section 3.2, we use the source distribution to produce samples from several "low-degree distributions" of the form $f_j'(U_q)$, where the f_j's are low-degree polynomials of odd degree. We then apply the function Tr on each sample. We make sure that the (f_j')s have the property that the sum of any subset of them is also a polynomial f' of odd degree. We use this property together with the additivity of Tr to show that the parity of any subset of the output bits is close to uniform. We then conclude using Lemma 3.6.

Lemma 3.7. *Let q be a power of 2. Fix any integers d and n. For every $j \in [d]$, define the multivariate polynomial $f_j : \mathbb{F}_q^n \to \mathbb{F}_q$ by $f_j(x) = \sum_{i=1}^{n} x_i^{2^j+(2i-1)}$. The function $D : \mathbb{F}_q^n \to \{0,1\}^d$ defined by $D_j(x) = Tr(f_j(x))$ is a deterministic $(1,\epsilon)$-affine source extractor, where $\epsilon = \frac{(d+n)\cdot 2^{d/2}}{\sqrt{q}}$.*

Proof. Fix an $(n,1)_q$-affine source $X \sim [t \cdot a + b]_{t \leftarrow U_q}$. Fix a nonempty subset $S \subseteq [d]$. We have

$$\bigoplus_{j \in S} D_j(X) \sim \bigoplus_{j \in S} Tr(f_j(X))$$

$$\sim Tr\left(\sum_{j \in S} f_j(X)\right)$$

$$\sim \left[Tr\left(\sum_{j \in S}\sum_{i=1}^{n}(t \cdot a_i + b_i)^{2j+(2i-1)}\right)\right]_{t \leftarrow U_q}.$$

Denote $f'(t) = \sum_{j \in S}\sum_{i=1}^{n}(t \cdot a_i + b_i)^{2j+(2i-1)}$. Note that f' is a polynomial of odd degree m where $m \leq 2d + 2n$. Therefore, using Corollary 3.5 we have

$$\bigoplus_{j \in S} D_j(X) \sim Tr(f'(U_q)) \overset{\frac{d+n}{\sqrt{q}}}{\approx} U_1.$$

Using Lemma 3.6 we get

$$|D(X) - U_d| \leq \frac{(d+n) \cdot 2^{d/2}}{\sqrt{q}}.$$

□

We now deal with fields of odd size. The proof is roughly analogous to the case of even-sized fields but requires a bit more work.

We will need the following special case of a lemma from [59].

Lemma 3.8. *[59][Lemma 4B, page 51] Let $q = p^l$ for some integer l and odd prime p. Let $f(t)$ be a polynomial in $\mathbb{F}_q[t]$. The following are equivalent.*

- *$f(t)$ is a square multiple in $\mathbb{F}_q[t]$.*
- *$f(t) = c \cdot (t - \nu_1)^{e_1} \cdots (t - \nu_s)^{e_s}$ for some $\nu_1, \ldots, \nu_s \in \overline{\mathbb{F}}_q$ and $c \in \mathbb{F}_q$, where e_i is even for all $i \in [s]$.*

Lemma 3.9. *Let $q = p^l$ for some integer l and odd prime p. Fix any integers d and n such that $d \leq q$. Let $c_1, \ldots, c_d$ be distinct elements in $\mathbb{F}_q$. Define the multivariate polynomial $f_0 : \mathbb{F}_q^n \to \mathbb{F}_q$ by $f_0(x) = \sum_{i=1}^{n} x_i^{2i-1}$. For $j \in [d]$, define the multivariate polynomial $f_j : \mathbb{F}_q^n \to \mathbb{F}_q$ by $f_j(x) = f_0(x) + c_j$. The function $D : \mathbb{F}_q^n \to \{0,1\}^d$ defined by $D_j(x) = QR(f_j(x))$ is a deterministic $(1,\epsilon)$-affine source extractor, where $\epsilon = \frac{4dn \cdot 2^{d/2}}{\sqrt{q}}$.*

Proof. Fix an $(n,1)_q$-affine source $X \sim [t \cdot a + b]_{t \leftarrow U_q}$. Fix a nonempty subset $S \subseteq [d]$. For any $x = t \cdot a + b$ in $Supp(X)$, we have

$$\bigoplus_{j\in S} D_j(x) = \bigoplus_{j\in S} QR(f_j(x))$$

$$= \bigoplus_{j\in S} QR\left(\left(\sum_{i=1}^{n}(t\cdot a_i + b_i)^{2i-1}\right) + c_j\right)$$

For $j \in S$, denote $f'_j(t) = \left(\sum_{i=1}^{n}(t\cdot a_i + b_i)^{2i-1}\right) + c_j$. For $x = t\cdot a + b$, we call x *good* if $f'_j(t) \neq 0$ for every $j \in S$. For any good $x = t \cdot a + b$, we have

$$\bigoplus_{j\in S} D_j(x) = \bigoplus_{j\in S} QR(f'_j(t)) = QR\left(\prod_{j\in S} f'_j(t)\right)$$

Since there are at most $d \cdot 2n$ bad xs, we get

$$\left|\bigoplus_{j\in S} D_j(X) - QR\left(\prod_{j\in S} f'_j(U_q)\right)\right| \leq d\cdot 2n/q.$$

Denote $f'(t) = \prod_{j\in S} f'_j(t)$. We will show that $f'(t)$ is not a square multiple in $\mathbb{F}_q[t]$. Fix some $j_0 \in S$. Since f'_{j_0} has odd degree it is not a square multiple in $\mathbb{F}_q[t]$. Therefore, by Lemma 3.8 (and by the fact that any polynomial decomposes into linear factors in $\overline{\mathbb{F}}_q$), $f'_{j_0}(t) = c\cdot(t-\nu_1)^{e_1}\cdots(t-\nu_s)^{e_s}$ for distinct $\nu_1,\ldots,\nu_s \in \overline{\mathbb{F}}_q$, where e_k is odd for some $k \in [s]$. Assuming that $|S| \geq 2$, fix any $j_1 \in S$ where $j_1 \neq j_0$. For any $t \in \overline{\mathbb{F}}_q$, $f'_{j_0}(t) - f'_{j_1}(t) = c_{j_0} - c_{j_1} \neq 0$. Therefore, f'_{j_0} and f'_{j_1} do not have a common linear factor in $\overline{\mathbb{F}}_q$. Hence, the factor $(t-\nu_k)$ appears an odd number of times in $f'(t) = \prod_{j\in S} f'_j(t)$. Therefore, by Lemma 3.8 $f'(t)$ is not a square multiple in $\mathbb{F}_q$. Thus, using Corollary 3.6 we have

$$\left|\bigoplus_{j\in S} D_j(X) - U_1\right| \leq \left|\bigoplus_{j\in S} D_j(X) - QR\left(f'(U_q)\right)\right| + \left|QR\left(f'(U_q)\right) - U_1\right|$$

$$\leq \frac{d\cdot 2n}{q} + \frac{2dn}{\sqrt{q}} \leq \frac{4dn}{\sqrt{q}}.$$

Therefore, using Lemma 3.6 we have

$$|D(X) - U_d| \leq \frac{4dn\cdot 2^{d/2}}{\sqrt{q}}.$$

□

We restate and prove Theorem 3.2

Theorem 3.2 For any field $\mathbb{F}_q$, integer n and $\epsilon > 0$, there is an explicit deterministic $(1,\epsilon)$-affine source extractor $D : \mathbb{F}_q^n \to \{0,1\}^d$, with $d = \lfloor \log q - 2\log(n/\epsilon) - 2\log\log q - 4 \rfloor$.

Proof. Using Lemmas 3.7 and 3.9, we can get an explicit deterministic $(1,\epsilon)$-affine source extractor $D : \mathbb{F}_q^n \to \{0,1\}^d$ such that

$$\epsilon \le \frac{4dn \cdot 2^{d/2}}{\sqrt{q}}.$$

Squaring, we get

$$\epsilon^2 \le \frac{16d^2n^2 \cdot 2^d}{q}.$$

Taking the logarithm on both sides, we get

$$2\log(\epsilon) \le 4 + 2\log d + 2\log n + d - \log q.$$

Rearranging and using $d \le \log q$, we get

$$d \ge \log q - 2\log(n/\epsilon) - 2\log\log q - 4.$$

□

We also prove the following instantiation of Lemmas 3.7 and 3.9, which we will use in the proof of Theorem 3.1. The following lemma states that we can extract any constant fraction of the randomness from an $(n,1)_q$-affine source, provided q is a large enough polynomial in n.

Lemma 3.10. *Fix any constant $0 < \delta < 1$. There exists a constant q_0 (depending on δ) such that for any prime power q and integer n with $q > q_0$ and $q \ge n^{7/\delta}$, there is an explicit deterministic $(1,\epsilon)$-affine source extractor $D : \mathbb{F}_q^n \to \{0,1\}^d$ where $\epsilon \le q^{-\delta/3}$ and $d = \lfloor (1-\delta)\log q \rfloor$.*

Proof. According to whether q is even or odd we use Lemma 3.7 or 3.9 with d and n as stated in the lemma. We get an explicit deterministic $(1,\epsilon)$-affine source extractor $D : \mathbb{F}_q^n \to \{0,1\}^d$ where

$$\epsilon \le \frac{4dn \cdot 2^{d/2}}{\sqrt{q}} \le \frac{4 \cdot (1-\delta)\log q \cdot q^{\delta/7} \cdot q^{\frac{1-\delta}{2}}}{\sqrt{q}}.$$

We take q large enough so that $q^{\delta/42} \ge 4 \cdot (1-\delta)\log q$. For such q, we have

$$\epsilon \le \frac{q^{\delta/42+\delta/7+1/2-\delta/2}}{q^{1/2}} = q^{-\delta/3}.$$

□

3.6 A Linear Seeded Extractor for Affine Sources

In this section we describe our construction of linear seeded affine source extractors. As described in Section 3.2, this seeded extractor will be used as a component in our construction of deterministic affine source extractors.

Given $u \in \mathbb{F}_q$ and an integer k, we define a $k \times n$ matrix $T_{u,k}$ by $(T_{u,k})_{j,i} = u^{ji}$ (where ji is an integer product). That is,

$$T_{u,k}(x) = \left(\sum_{i=1}^{n} x_i \cdot u^i, \sum_{i=1}^{n} x_i \cdot u^{2i}, \ldots, \sum_{i=1}^{n} x_i \cdot u^{ki} \right)$$

for $x \in \mathbb{F}_q^n$.

The following theorem shows how to extract all the randomness from an $(n,k)_q$-affine source using a seed of length $\lceil \log n + 2\log k + \log(1/\epsilon) \rceil$ whenever $q > 2n \cdot k^2/\epsilon$.

Theorem 3.7. *Fix any field $\mathbb{F}_q$, integers n, k, and $\epsilon > 0$ such that $q \geq 2\frac{n \cdot k^2}{\epsilon}$. Let s be the smallest power of 2 such that $s \geq \frac{n \cdot k^2}{\epsilon}$. Let $U = \{u_1, \ldots, u_s\}$ be a set of distinct elements in $\mathbb{F}_q$. Let $d = \log s$. We identify $[s]$ with $\{0,1\}^d$. The function $E : \mathbb{F}_q^n \times \{0,1\}^d \to \mathbb{F}_q^k$ defined by*

$$E(x,y) = T_{u_y,k}(x) = \left(\sum_{i=1}^{n} x_i \cdot u_y^i, \sum_{i=1}^{n} x_i \cdot u_y^{2i}, \ldots, \sum_{i=1}^{n} x_i \cdot u_y^{ki} \right)$$

is a linear seeded (k,ϵ)-affine source extractor.

The theorem will be derived easily from the following lemma.

Lemma 3.11. *Fix any field $\mathbb{F}_q$ and integers n, k such that $q \geq n \cdot k^2$. Fix any affine subspace $A \subseteq \mathbb{F}_q^n$ of dimension k. There are at most $n \cdot k^2$ elements $u \in \mathbb{F}_q$ such that $T_{u,k}(A) \subsetneq \mathbb{F}_q^k$.*

Proof. We denote $T_u = T_{u,k}$. First note that if $A = A_1 + b$ where $b \in \mathbb{F}_q^n$ and A_1 is a linear subspace of dimension k, then $(T_u(A_1) = \mathbb{F}_q^k) \leftrightarrow (T_u(A) = \mathbb{F}_q^k)$. Therefore, we assume A is a linear subspace with basis $\{a^{(1)}, a^{(2)}, \ldots, a^{(k)}\}$ where $a^{(j)} \in \mathbb{F}_q^n$. Denote by B the $n \times k$ matrix

$$B = \left(a^{(1)}, a^{(2)}, \ldots, a^{(k)} \right).$$

We have

$$T_u(A) = T_u \cdot B(\mathbb{F}_q^k)$$

where $\cdot$ denotes the matrix product.

Denote by C_u the $k \times k$ matrix $T_u \cdot B$. That is,

$$(C_u)_{j,l} = \sum_{i=1}^{n} a^{(l)}{}_i \cdot u^{ji}.$$

$$C_u = \begin{pmatrix} \sum_{i=1}^n a^{(1)}{}_i \cdot u^i & \sum_{i=1}^n a^{(2)}{}_i \cdot u^i & \dots & \sum_{i=1}^n a^{(k)}{}_i \cdot u^i \\ \sum_{i=1}^n a^{(1)}{}_i \cdot u^{2i} & \sum_{i=1}^n a^{(2)}{}_i \cdot u^{2i} & \dots & \sum_{i=1}^n a^{(k)}{}_i \cdot u^{2i} \\ . & . & \dots & . \\ . & . & \dots & . \\ . & . & \dots & . \\ \sum_{i=1}^n a^{(1)}{}_i \cdot u^{ki} & \sum_{i=1}^n a^{(2)}{}_i \cdot u^{ki} & \dots & \sum_{i=1}^n a^{(k)}{}_i \cdot u^{ki} \end{pmatrix}$$

Recall that $(C_u(\mathbb{F}_q^k) = \mathbb{F}_q^k) \leftrightarrow (Det(C_u) \neq 0)$.

Let $f(u) = Det(C_u)$. We will show that $f(u)$ is a nonzero polynomial of degree at most $n \cdot k^2$. It follows that $Det(C_u) = 0$ for at most $n \cdot k^2$ us and the lemma follows.

$$f(u) = Det(C_u) = \sum_{\sigma \in S_k} sgn(\sigma) \cdot f_\sigma(u)$$

where

$$f_\sigma(u) = \prod_{j=1}^{k} (C_u)_{j,\sigma(j)}.$$

For $j \in [k]$, we define $j_{\max}$ to be the maximal $i \in [n]$ such that $a^{(j)}{}_i$ is nonzero. Note that, using Gaussian elimination, we can find a basis $a^{(1)}, \dots, a^{(k)}$ of A such that,

$$0 < 1_{\max} < 2_{\max} < \dots < k_{\max}.$$

We assume without loss of generality that this is the case. Let $Id \in S_k$ be the identity permutation. We will show that for every $\sigma \neq Id$ in S_k, $deg(f_\sigma) < deg(f_{Id})$.

Assume for contradiction that there exists $\sigma \neq Id$ in S_k with $deg(f_\sigma) \geq deg(f_{Id})$. Fix such a permutation σ that maximizes $\deg(f_\sigma)$. (That is, $deg(f_\sigma) \geq deg(f_{\sigma'})$ for every $\sigma' \in S_k$). $(C_u)_{j,\sigma(j)}$ is a polynomial in u of degree $j \cdot \sigma(j)_{max}$. Therefore, $f_\sigma(u)$ has degree $\sum_{j=1}^k j \cdot \sigma(j)_{max}$. Since $\sigma \neq Id$, there exist $j_1 < j_2$ such that $\sigma(j_1) > \sigma(j_2)$. Let $\tau = (\sigma(j_1)\sigma(j_2)) \cdot \sigma$, i.e., the permutation τ consists of applying σ and then "switching" between $\sigma(j_1)$ and $\sigma(j_2)$.

We have

$$\begin{aligned} deg(f_\tau) - deg(f_\sigma) &= j_2(\sigma(j_1)_{max} - \sigma(j_2)_{max}) + j_1(\sigma(j_2)_{max} - \sigma(j_1)_{max}) \\ &= j_2(\sigma(j_1)_{max} - \sigma(j_2)_{max}) - j_1(\sigma(j_1)_{max} - \sigma(j_2)_{max}) \\ &= (j_2 - j_1)(\sigma(j_1)_{max} - \sigma(j_2)_{max}) > 0, \end{aligned}$$

which contradicts the maximality of $deg(f_\sigma)$.

Therefore, for any $\sigma \neq Id$, we have $deg(f_{Id}) > deg(f_\sigma)$. Thus, f_{Id} cannot be "canceled out" by the other summands in $f(u)$, and $f(u)$ is a nonzero polynomial of degree $deg(f_{Id}) = \sum_{j=1}^{k} j \cdot j_{max} \leq n \cdot \sum_{j=1}^{k} j = n \cdot \frac{k(k+1)}{2} \leq n \cdot k^2$. □

We can now easily prove the theorem.

Proof. (of Theorem 3.7) Fix any $(n,k)_q$-affine source X. Using Lemma 3.11 we get

$$\Pr_{y \leftarrow U_d}(E(X,y) \nsim U_{\mathbb{F}_q^k}) \leq \frac{n \cdot k^2}{|U|} \leq \epsilon.$$

Therefore, by Lemma 3.3

$$E(X, U_d) \overset{\epsilon}{\sim} U_{\mathbb{F}_q^k}.$$

□

Remark 3.12. *Actually, Lemma 3.11 implies that the extractor E from Theorem 3.7 is* strong. *That is, the distribution $(U_d, E(X, U_d))$ is also close to uniform.*

3.7 Composing Extractors

Let E be a linear seeded affine source extractor. In this section, we show that we can use E with a correlated seed that we have extracted deterministically from our affine source. The argument we use is based on the recycling paradigm described in the first chapter.

Our starting point will be the following lemma, which is a combination of Lemmas 2.5 and 2.6 in [26].[12] Fix a distribution X on $\mathbb{F}_q^n$ and functions T and D. Roughly speaking, the lemma states that if $D(X)$ is close to uniform even when conditioning on a certain output value of T, then the output distribution $T(X)$ is "almost not affected" by conditioning on a value of D.

Lemma 3.13 ([26]). *Let X be a distribution on $\mathbb{F}_q^n$. Let $T : \mathbb{F}_q^n \to \mathbb{F}_q^m$ and $D : \mathbb{F}_q^n \to \{0,1\}^d$ be any functions. Assume that for every $a \in Supp(T(X))$ we have $|(D(X)|T(x) = a) - U_d| \leq \epsilon$. Then for every $y \in Supp(D(X))$ we have*

$$(T(X)|D(x) = y) \overset{\epsilon \cdot 2^{d+1}}{\sim} T(X).$$

The following corollary of Lemma 3.13 shows that, for a fixed linear mapping T, the output distribution of T on an affine source X is "almost not affected" by conditioning on an output value of a deterministic affine source extractor D.

[12] In [26] the authors assume all distributions are over binary strings, but it is easy to see that the proof follows in the case stated here.

Corollary 3.8. *Fix any field $\mathbb{F}_q$, integers n, k, m, d, and $\epsilon > 0$ such that $k > m$ and $\epsilon < 2^{-(d+1)}$. Let $D : \mathbb{F}_q^n \to \{0,1\}^d$ be a deterministic $(1, \epsilon)$-affine source extractor. Let X be an $(n, k)_q$-affine source. Then for any linear mapping $T : \mathbb{F}_q^n \to \mathbb{F}_q^m$ and $y \in \{0,1\}^d$, we have*

$$|(T(X)|D(x) = y) - T(X)| \leq \epsilon \cdot 2^{d+1}.$$

Proof. Fix any $a \in Supp(T(X))$. It is easy to see that $(X|T(x) = a)$ is an $(n, k')_q$-affine source for some $k' \geq 1$ (since $k > m$). Therefore,

$$(D(X)|T(x) = a) \overset{\epsilon}{\sim} U_d.$$

Fix any $y \in \{0,1\}^d$. Since $\epsilon < 2^{-d}$, we know that $y \in Supp(D(X))$. Thus, using lemma 3.13, we have

$$|(T(X)|D(x) = y) - T(X)| \leq \epsilon \cdot 2^{d+1}.$$

□

Corollary 3.8 works for any output value y and linear mapping T. In particular, as observed in [26], it will work for an output value y and linear mapping T_y that is determined by y. We use this fact to compose a deterministic affine source extractor with a linear seeded affine source extractor, and get a new deterministic affine source extractor that extracts more randomness.

Theorem 3.9. *Fix any field $\mathbb{F}_q$, integers n, k, m, d, and $\epsilon, \epsilon' > 0$, such that $k > m$ and $\epsilon' < 2^{-(d+1)}$. Let $D' : \mathbb{F}_q^n \to \{0,1\}^d$ be a deterministic $(1, \epsilon')$-affine source extractor. Let $E : \mathbb{F}_q^n \times \{0,1\}^d \to \mathbb{F}_q^m$ be a linear seeded (k, ϵ)-affine source extractor. Then $D : \mathbb{F}_q^n \to \mathbb{F}_q^m$ defined by*

$$D(x) = E(x, D'(x))$$

is a deterministic (k, ρ)-affine source extractor, where $\rho = 4\epsilon' \cdot 2^d + \epsilon$.

Proof. Fix an $(n, k)_q$-affine source X. Note that,

$$D(X) \sim E(X, D'(X)) \sim \sum_{y \in \{0,1\}^d} \Pr(D'(X) = y) \cdot (E(X, y)|D'(x) = y),$$

and

$$E(X, U_d) \sim \sum_{y \in \{0,1\}^d} \Pr(U_d = y) \cdot E(X, y).$$

We know that $|D'(X) - U_d| \leq \epsilon'$. Fix any $y \in \{0,1\}^d$. $T_y(x) \triangleq E(x, y)$ is a linear mapping from $\mathbb{F}_q^n$ to $\mathbb{F}_q^m$, where $m < k$. Therefore, by Corollary 3.8, we have

$$|(E(X, y)|D'(x) = y) - E(X, y)| \leq \epsilon' \cdot 2^{d+1}.$$

By Lemma 3.2, we have

$$|D(X) - E(X, U_d)| \leq 2\epsilon' + \epsilon' \cdot 2^{d+1}.$$

Therefore,

$$|D(X) - U_{\mathbb{F}_q^m}| \leq 2\epsilon' + \epsilon' \cdot 2^{d+1} + \epsilon \leq 4\epsilon' \cdot 2^d + \epsilon.$$

□

3.8 Putting It All Together

In this section we present our main extractor construction.

Using Theorem 3.9, we compose the deterministic extractor of Lemma 3.10 and the seeded extractor of Theorem 3.7 to get a deterministic extractor that extracts almost all the randomness from an $(n, k)_q$-affine source assuming q is a large enough polynomial in n. We restate and prove Theorem 3.1.

Theorem 3.1 There exists a constant q_0 such that for any field $\mathbb{F}_q$ and integers n, k with $q > \max[q_0, n^{20}]$, there is an explicit deterministic (k, ρ)-affine source extractor $D : \mathbb{F}_q^n \to \mathbb{F}_q^{k-1}$, with $\rho \leq q^{-1/21}$.

Proof. We use Lemma 3.10 with $\delta = 4/5$. For large enough q and any $n \leq q^{\delta/7}$, we get an explicit deterministic $(1, \epsilon')$-affine source extractor $D' : \mathbb{F}_q^n \to \{0,1\}^{d'}$, where $d' = \lfloor (1/5) \log q \rfloor$ and $\epsilon' \leq q^{-4/15}$. We use Theorem 3.7 with parameters $q, n, k-1$ and $\epsilon = \frac{8n^3}{q^{1/5}}$. Note that,

$$\frac{2n \cdot k^2}{\epsilon} \leq \frac{2n^3 \cdot q^{1/5}}{8n^3} \leq q,$$

as required in Theorem 3.7. We get a linear seeded (k, ϵ)-affine source extractor $E : \mathbb{F}_q^n \times \{0,1\}^d \to \mathbb{F}_q^{k-1}$, where $2^d \leq \frac{2n \cdot k^2}{\epsilon} \leq q^{1/5}/4 \leq 2^{d'}$. Since $d \leq d'$, we can use Theorem 3.9 with D' and E and get an explicit deterministic (k, ρ)-affine source extractor $D : \mathbb{F}_q^n \to \mathbb{F}_q^{k-1}$, where

$$\rho = 4\epsilon' \cdot 2^d + \epsilon \leq 4q^{-4/15} \cdot q^{1/5}/4 + \frac{8n^3}{q^{1/5}}$$

$$\leq q^{-1/15} + 8 \cdot q^{3/20-1/5} \leq 9 \cdot q^{-1/20} \leq q^{-1/21}$$

for large enough q. □

Chapter 4

Extractors and Rank Extractors for Polynomial Sources

Summary

In this chapter we construct explicit deterministic extractors from *polynomial sources*, namely from distributions sampled by low-degree multivariate polynomials over finite fields. This naturally generalizes previous work on extraction from affine sources (which are degree 1 polynomials). A direct consequence is a deterministic extractor for distributions sampled by polynomial-size arithmetic circuits over exponentially large fields.

The steps in our extractor construction, and the tools (mainly from algebraic geometry) that we use for them, are of independent interest.

The first step is a construction of *rank extractors*, which are polynomial mappings that "extract" the algebraic rank from any system of low-degree polynomials. More precisely, for any n polynomials, k of which are algebraically independent, a rank extractor outputs k algebraically independent polynomials of slightly higher degree. The rank extractors we construct are applicable not only over finite fields but also over fields of characteristic zero.

The next step is relating algebraic independence to min-entropy. We use a theorem of Wooley to show that these parameters are tightly connected. This allows replacing the algebraic assumption on the source (above) by the natural information-theoretic one. It also shows that a rank extractor is already a high-quality *condenser* for polynomial sources over polynomially large fields.

A. Gabizon, *Deterministic Extraction from Weak Random Sources*,
Monographs in Theoretical Computer Science. An EATCS Series,
DOI 10.1007/978-3-642-14903-0_4, © Springer-Verlag Berlin Heidelberg 2011

Finally, to turn the condensers into extractors, we employ a theorem of Bombieri, giving a character sum estimate for polynomials defined over curves. It allows extracting all the randomness (up to a multiplicative constant) from polynomial sources over exponentially large fields.

4.1 Introduction

A natural generalization of affine sources is allowing sources that can be sampled by low-degree multivariate polynomials. Let $\mathbb{F}$ be a field (finite or infinite). For integers $k \leq n$ and d we consider the family of all mappings $x : \mathbb{F}^k \to \mathbb{F}^n$ that are defined by polynomials of total degree at most d (we denote our mapping by x since this will represent our source). That is,

$$x(t) = (x_1(t_1, \ldots, t_k), \ldots, x_n(t_1, \ldots, t_k)),$$

where for each $1 \leq i \leq n$ the coordinate x_i of the mapping is a k-variate polynomial of total degree at most d. We denote this set of mappings by $\mathcal{M}(\mathbb{F}^k \to \mathbb{F}^n, d)$. We will focus on the case where the field $\mathbb{F}$ is much larger than d and will specify in each result how large the field has to be. This will allow us to refer to the elements of $\mathcal{M}(\mathbb{F}^k \to \mathbb{F}^n, d)$ as *low-degree* mappings. It is important to note that any weak source can be represented as the image of *some* polynomial mapping over a finite field $\mathbb{F}$. However, in general, the polynomials representing the source will have very high degrees (this can be seen by a simple counting argument). Since it is known [47] that deterministic extraction from arbitrary sources is impossible, we see that restricting our attention to low-degree mappings is essential.

For affine sources we have the requirement that the affine mapping defining the source be non-degenerate. This ensures that the source sampled by this mapping has "enough" entropy. We would like to extend this requirement also to the case of low-degree mappings in $\mathcal{M}(\mathbb{F}^k \to \mathbb{F}^n, d)$. The way to generalize this notion is via the partial derivative matrix (sometimes called the *Jacobian*) of a mapping $x \in \mathcal{M}(\mathbb{F}^k \to \mathbb{F}^n, d)$. This is an $n \times k$ matrix denoted by $\frac{\partial x}{\partial t}$, defined as follows:

$$\frac{\partial x}{\partial t} \triangleq \begin{pmatrix} \frac{\partial x_1}{\partial t_1} & \cdots & \frac{\partial x_1}{\partial t_k} \\ \vdots & \ddots & \vdots \\ \frac{\partial x_n}{\partial t_1} & \cdots & \frac{\partial x_n}{\partial t_k} \end{pmatrix}$$

where the partial derivatives are defined in the standard way, as formal derivatives of polynomials. Let us define the *rank* of $x \in \mathcal{M}(\mathbb{F}^k \to \mathbb{F}^n, d)$ to be the rank of the matrix $\frac{\partial x}{\partial t}$ when considered as a matrix over the field of rational functions in variables $t_1, \ldots, t_k$. We say that $x \in \mathcal{M}(\mathbb{F}^k \to \mathbb{F}^n, d)$ is *non-degenerate* if its rank is k (obviously, x cannot have rank larger than k).

Definition 4.1 (polynomial source). *Let $\mathbb{F}$ be a finite field. A distribution X over $\mathbb{F}^n$ is an (n,k,d)-*polynomial source *over $\mathbb{F}$ if there exists a non-degenerate mapping $x \in \mathcal{M}(\mathbb{F}^k \to \mathbb{F}^n, d)$ such that X is sampled by choosing t uniformly at random in $\mathbb{F}^k$ and outputting $x(t)$.*

It is easy to see that the above definition of a polynomial source is indeed a generalization of the affine case, since the partial derivative matrix of an affine mapping is simply its coefficient matrix (in some basis). We note that while low-degree polynomials play an essential role in complexity theory, extraction from sources defined by such polynomials has apparently not been studied before.

Rank and min-entropy: One reason for using the rank of the partial derivative matrix is that, over sufficiently large prime fields, it allows us to prove a lower-bound on the entropy of an (n,k,d)-polynomial source. This lower bound follows from a theorem of Wooley [73] (see Theorem 4.1). Roughly speaking, Wooley's theorem implies that a distribution sampled by a non-degenerate mapping $x \in \mathcal{M}(\mathbb{F}^k \to \mathbb{F}^n, d)$ is close (in statistical distance) to a distribution with min-entropy at least $k \cdot \log\left(\frac{|\mathbb{F}|}{2d}\right)$. Rewriting this quantity as

$$\left(1 - \frac{\log(2d)}{\log(|\mathbb{F}|)}\right) \cdot k \cdot \log(|\mathbb{F}|),$$

we see the way in which, as $|\mathbb{F}|$ grows, this bound "approaches" the entropy bound of $k \cdot \log(|\mathbb{F}|)$ we have for affine sources of the same rank.

Rank and algebraic independence: Over fields of exponential characteristic (or of characteristic zero) we will see that the above notion of the rank of a mapping coincides with the more intuitive notion of *algebraic independence* (see Section 4.2 for the relevant definitions). Roughly speaking, over such fields, a mapping $x = (x_1, \ldots, x_n) \in \mathcal{M}(\mathbb{F}^k \to \mathbb{F}^n, d)$ has rank k iff the set of polynomials $\{x_1(t), \ldots, x_n(t)\}$ contains k algebraically independent polynomials (we should note that the direction "rank $k \to$ algebraic independence" is true over any field, regardless of its characteristic). Since we want some of our results to hold also over fields of polynomial size we opt to use the rank of the partial derivative matrix in our definition of a polynomial source. In Section 4.3 we give a detailed discussion of the connection between algebraic independence and rank. Our proofs are direct extensions of the treatment appearing in [21] and [43], where the equivalence between the two notions is shown over the complex numbers.

4.1.1 Rank Extractors

The above discussion of polynomial sources raises the following natural question: Can we "extract" the rank of these sources without destroying their structure? In other words, can we construct a *fixed* polynomial mapping $y : \mathbb{F}^n \to \mathbb{F}^k$ such that for any non-degenerate $x \in \mathcal{M}(\mathbb{F}^k \to \mathbb{F}^n, d)$ the composition of y with x is a non-degenerate mapping from $\mathbb{F}^k$ to $\mathbb{F}^k$? We call a non-degenerate mapping $z : \mathbb{F}^k \to \mathbb{F}^k$ a *full rank* mapping and a mapping y satisfying the above condition a *rank extractor.*

Definition 4.2 (rank extractor). *Let $\mathbb{F}$ be some field. Let $y : \mathbb{F}^n \to \mathbb{F}^k$ be a polynomial mapping defined by*

$$y(x) = (y_1(x_1, \dots, x_n), \dots, y_k(x_1, \dots, x_n)),$$

where each y_i is a multivariate polynomial over $\mathbb{F}$. We say that y is an (n, k, d)-rank extractor over $\mathbb{F}$ if for every non-degenerate mapping $x \in \mathcal{M}(\mathbb{F}^k \to \mathbb{F}^n, d)$ the composition $y \circ x : \mathbb{F}^k \to \mathbb{F}^k$ has rank k. We will call such a mapping y explicit *if it can be computed in polynomial time.*

Clearly, a construction of a rank extractor will bring us closer to constructing an extractor for low-degree polynomial sources. Using an explicit rank extractor reduces the problem of constructing an extractor for arbitrary polynomial sources into the problem of constructing an extractor for polynomial sources of full rank. Surprisingly enough, the problem of extraction from full rank sources is not so easy and requires the use of deep results from algebraic geometry.

Our first main result is a construction of an explicit (n, k, d)-rank extractor over $\mathbb{F}$, where $\mathbb{F}$ can be any field of characteristic zero or of characteristic at least poly(n, d). It is natural to require that the degree of the rank extractor be as small as possible. Clearly the degree has to be larger than 1 since an affine mapping cannot be a rank extractor (because we can always "hide" a polynomial source in the kernel of such a mapping). The rank extractors we construct have degree that is bounded by a polynomial in n and d. In Section 4.4 we prove the following theorem:

Theorem 4.1. *Let $k \le n$ and d be integers. Let $\mathbb{F}$ be a field of characteristic zero or of characteristic larger than $8k^2d^3n$. Then there exists an explicit (n, k, d)-rank extractor over $\mathbb{F}$ whose degree is bounded by $8k^2d^2n$. Moreover, this rank extractor can be computed in time poly$(n, \log(d))$.*

We note that our construction of rank extractors does not depend on the underlying field. We give a single construction, defined using integers, that is a rank extractor over any field satisfying the conditions of Theorem 4.1.

4.1.2 Extractors and Condensers for Polynomial Sources

As was mentioned in the previous section, applying the rank extractor given by Theorem 4.1 reduces the problem of constructing an extractor for (n, k, d)-polynomials sources into the problem of constructing an extractor for (k, k, d')-polynomial sources, where d' is the degree of the source obtained *after* applying the rank extractor (since Theorem 4.1 implies that d' is polynomial in n and d). Our second main result is a construction of such an extractor. Before stating our result we give a formal definition of an extractor for polynomial sources.

Definition 4.3 (extractor). *Let $k \leq n$ and d be integers. Let $\mathbb{F}$ be a finite field. A function $E : \mathbb{F}^n \to \{0,1\}^m$ is a (k, d, ϵ)-*extractor for polynomial sources *if for every (n, k, d)-polynomial source X over $\mathbb{F}^n$, the random variable $E(X)$ is ϵ-close to the uniform distribution on $\{0,1\}^m$. We say that E is* explicit *if it can be computed in poly$(n, \log(d))$ time.*

The following theorem, proved in Section 4.5, asserts the existence of an explicit extractor for full-rank polynomial sources over sufficiently large prime fields. The output length of this extractor is $\Omega(k \cdot \log(|\mathbb{F}|))$, which is within a multiplicative constant of the maximal length possible. The main tool in the proof of our theorem is a theorem of Bombieri [8] giving exponential sum estimates for polynomials defined over low-degree curves.

Theorem 4.2. *There exist absolute constants C and c such that the following holds: Let k and $d > 1$ be integers and let $\mathbb{F}$ be a field of prime cardinality $p > d^{Ck}$. Then, there exists a function $E : \mathbb{F}^k \to \{0,1\}^m$ that is an explicit (k, d, ϵ)-extractor for polynomial sources over $\mathbb{F}^k$ with $m = \lfloor c \cdot k \cdot \log(p) \rfloor$ and $\epsilon = p^{-\Omega(1)}$.*

Combining Theorem 4.2 with Theorem 4.1 gives an extractor for general polynomial sources. This extractor, whose existence is stated in the following corollary, also has output length which is within a multiplicative constant of optimal.

Corollary 4.1. *There exist absolute constants C and c such that the following holds: Let $k \leq n$ and $d > 1$ be integers and let $d' = 8k^2d^3n$. Let $\mathbb{F}$ be a field of prime cardinality $p > (d')^{Ck}$. Then, there exists a function $E : \mathbb{F}^k \to \{0,1\}^m$ that is an explicit (k, d, ϵ)-extractor for polynomial sources over $\mathbb{F}^n$ with $m = \lfloor c \cdot k \cdot \log(p) \rfloor$ and $\epsilon = p^{-\Omega(1)}$.*

It is possible to improve the output length of our extractors so that it is equal to a $(1 - \alpha)$-fraction of the source min entropy, for any constant $\alpha > 0$. This improvement, which was suggested to us by Salil Vadhan, is described in Section 4.6.

We note that both in the last corollary and in Theorem 4.2, the bound on the field size does not pose a computational problem. Over a finite field $\mathbb{F}$, arithmetic operations can be performed in time polynomial in $\log(|\mathbb{F}|)$, and hence all computations required by the extractor can be performed in polynomial time. However, it remains an interesting open problem whether extraction can be performed over smaller fields, say of size polynomial in n and d.

Condensers over polynomially large fields: We note that over polynomially large fields, our techniques give a deterministic *condenser* for polynomial sources. A condenser is a relaxation of an extractor and is required to output a distribution with "high" min-entropy rather than a uniform distribution. The word "condenser" implies that the length of the output should be smaller than the length of the input. That is, the aim of a condenser is to "compress" the source while keeping as much of the entropy as possible. For convenience we define condensers as mappings over alphabet $\mathbb{F}$ rather than by the standard definition using binary alphabet.

Definition 4.4 (condenser). *Let $\mathcal{D}$ be a family of distributions over $\mathbb{F}^n$. A function $C : \mathbb{F}^n \to \mathbb{F}^m$ is an (ϵ, k')-*condenser *for $\mathcal{D}$ if for every X in $\mathcal{D}$ the distribution $C(X)$ is ϵ-close to having min-entropy at least k'. A condenser is* explicit *if it can be computed in polynomial time.*

From Wooley's theorem [73], mentioned earlier, it follows that if we apply a rank extractor to a polynomial source we get a source which is close to having high min-entropy. The next theorem follows immediately by combining Theorem 4.1 and Wooley's theorem (Corollary 4.3).

Theorem 4.3. *Let $k \leq n$ and d, d' be integers. Let $\mathbb{F}$ be a field of prime cardinality larger than $d \cdot d'$. Let $y : \mathbb{F}^n \to \mathbb{F}_q^k$ an (n, k, d)-rank extractor such that* $\deg(y) \leq d'$. *Then y is an (ϵ, k')-condenser for the family of (n, k, d)-polynomial sources over $\mathbb{F}$, where $\epsilon = \frac{d \cdot d' \cdot k}{|\mathbb{F}|}$ and $k' = k \cdot \log(|\mathbb{F}|/2dd')$.*

It should be noted that this condenser is "almost" the best one could hope for (without building an extractor, of course). To see this, suppose that $|\mathbb{F}| \approx (2d')^c$ for some constant $c > 1$. We get that the output of the condenser is close to having min-entropy

$$k' = k \cdot \log(|\mathbb{F}|/2d') \approx \left(1 - \frac{1}{c}\right) \cdot k \cdot \log(|\mathbb{F}|),$$

and so the ratio between the length of the output (in bits) and its min-entropy can be made arbitrarily close to 1 by choosing c to be large enough.

Dispersers over the complex field. A disperser is a relaxation of an extractor in which the output is only required to have large support (instead of being close to uniform). Dispersers are usually considered only for distributions over finite sets. However, for polynomial sources we can extend our view also for infinite sets (namely infinite fields). It is shown in [21] that the image of a full-rank mapping $x \in \mathcal{M}(\mathbb{C}^k \to \mathbb{C}^k, d)$ contains all of $\mathbb{C}^k$ except for the zero set of some polynomial. This shows that our rank extractors can be viewed as deterministic *dispersers* for polynomial sources over $\mathbb{C}$. That is, a rank extractor is a fixed polynomial transformation mapping *any* polynomial source into almost all of $\mathbb{C}^k$. We discuss this observation in Section 4.8.

4.1.3 Rank Versus Entropy — Weak Polynomial Sources

So far we focused on extraction from sources which were defined algebraically — we were given a bound on the algebraic rank of the set of polynomials we extract from. We now switch to the more standard definition (from the extractor literature standpoint) of extraction from sources with given min-entropy. These will be called *Weak Polynomial Sources.*

Definition 4.5 (weak polynomial source). *A distribution X over $\mathbb{F}^n$ is an (n, k, d)-weak polynomial source (WPS) if*

- *There exists a polynomial mapping $x \in \mathcal{M}(\mathbb{F}^n \to \mathbb{F}^n, d)$ such that X is sampled by choosing t uniformly in $\mathbb{F}^n$ and outputting $x(t)$.*
- *X has min-entropy at least $k \cdot \log(|\mathbb{F}|)$.*

Notice in the definition that the min-entropy threshold is $k \cdot \log(|\mathbb{F}|)$ (instead of just k). This is to hint at the connection (which we prove later) between the rank of the source and its entropy. Intuitively, a distribution sampled by a rank r mapping $x : \mathbb{F}^n \to \mathbb{F}^n$ "should" have entropy roughly $r \cdot \log(|\mathbb{F}|)$, and indeed, for affine sources, this is exactly the case.

The following theorem, whose proof can be found in Section 4.7, shows the existence of an explicit deterministic extractor for the class of weak polynomial sources.

Theorem 4.4. *There exist absolute constants C and c such that the following holds: Let $k \leq n$ and $d > 1$ be integers and let $d' = 8k^2d^3n$. Let $\mathbb{F}$ be a field of prime cardinality $p > (d')^{Ck}$. Then, there exists a function $E : \mathbb{F}^n \to \{0,1\}^m$ that is an explicit (k, d, ϵ)-extractor for weak polynomial sources over $\mathbb{F}^n$ with $m = \lfloor c \cdot k \cdot \log(p) \rfloor$ and $\epsilon = p^{-\Omega(1)}$.*

The parameters of the extractor given by the theorem can be seen to be roughly the same as those of the extractor for regular polynomial sources (Corollary 4.1). In fact, the extractor we use for weak polynomial sources is the same one we used for polynomial sources. The proof of Theorem 4.4 will follow by showing that any (n, k, d)-WPS is close (in statistical distance) to a convex combination of (n, k, d)-polynomial sources. This implies that any extractor that works for polynomial sources will work also for weak polynomial sources.

The entropy of a polynomial mapping. We can use the methods employed in the proof of Theorem 4.4 to show that over sufficiently large fields, the entropy of the output of a low-degree polynomial mapping $x \in \mathcal{M}(\mathbb{F}^n \to \mathbb{F}^n, d)$ is always "close" to $\text{rank}(x) \cdot \log(|\mathbb{F}|)$. This can be viewed as a generalization of the simple fact that for an *affine* mapping x, the entropy is always equal to $\text{rank}(x) \cdot \log(|\mathbb{F}|)$. (See Section 4.7.2 for the formal statement of this result.)

Extractors for poly-size arithmetic circuits. An interesting corollary of Theorem 4.4 is the existence of deterministic extractors for the class of distributions sampled by polynomial-size arithmetic circuits over exponentially large fields. This follows from the fact that the degrees of the polynomials computed by poly-size circuits are exponential, and the construction of an (n, k, d)-rank extractor is efficient even when d is exponential.

We say that a distribution X on $\mathbb{F}^n$ is sampled by a size s arithmetic circuit if there exists an arithmetic circuit A of size s with n inputs and n outputs such that the fan-in of each gate is at most 2 and such that X is the distribution of the output of A on a random input chosen uniformly from $\mathbb{F}^n$. We say that X is an (n, k, s)-*arithmetic source* if X is sampled by a size s arithmetic circuit and its min-entropy is at least $k \cdot \log(|\mathbb{F}|)$.

Corollary 4.2. *There exist absolute constants C and c such that the following holds: Let $k \leq n$ and $s > 1$ be integers. Let $d = 2^s$ and let $d' = 8k^2d^3n$. Let $\mathbb{F}$ be a field of prime cardinality $p > (d')^{Ck}$. Then, there exists an explicit function $E : \mathbb{F}^n \to \{0,1\}^m$ such that for every (n, k, s)-arithmetic source X over $\mathbb{F}$, the distribution of $E(X)$ is ϵ-close to uniform, where $m = \lfloor c \cdot k \cdot \log(p) \rfloor$ and $\epsilon = p^{-\Omega(1)}$. That is, E is an extractor for the class of (n, k, s)-arithmetic sources.*

It is interesting to contrast this result with the extractors of [68] from polynomial-size *boolean* circuits. Their extractors rely on complexity assumptions, and they prove that such assumptions are necessary. It is interesting that over large fields no such assumptions, nor lower bounds, are necessary.

4.1.4 Organization

Section 4.2 contains general preliminaries on probability distributions and finite field algebra. Section 4.3 contains a detailed discussion on the connection between algebraic independence and rank. In Section 4.4 we describe our construction of a rank extractor and prove Theorem 4.1. In Section 4.5 we construct and analyze an extractor for full-rank polynomial sources and prove Theorem 4.2. In Section 4.6 we show how to increase the output length of our extractors. In Section 4.7 we discuss extractors for weak polynomial sources and prove Theorem 4.4. In Section 4.8 we discuss rank extractors over the complex numbers. Appendix B contains background from algebraic geometry required for the proof of Theorem 4.2.

4.2 General Preliminaries

4.2.1 Probability Distributions

Let Ω be some finite set. Let P be a distribution on Ω. For $B \subseteq \Omega$, we denote $P(B)$, i.e., the probability of B according to P, by $\Pr_P(B)$ or $\Pr(P \subseteq B)$; When $B \in \Omega$, we will also use the notation $\Pr(P = B)$.

Given a function $A : \Omega \to U$, we denote by $A(P)$ the distribution induced on U when sampling t by P and calculating $A(t)$. When we write $t_1, \ldots, t_k \leftarrow P$, we mean that $t_1, \ldots, t_k$ are chosen *independently* according to P. We denote by U_Ω the uniform distribution on Ω. Given a function $x : \mathbb{F}^m \to \mathbb{F}$, we denote by $x(U_m)$ the distribution $x(U_{\mathbb{F}^m})$. For a distribution P on Ω^d and $j \in [d]$, we denote by P_j the restriction of P to the jth coordinate.

The *statistical distance* between two distributions P and Q on Ω, denoted by $|P - Q|$, is defined as

$$|P - Q| \triangleq \max_{S \subseteq \Omega} \left| \Pr_P(S) - \Pr_Q(S) \right| = \frac{1}{2} \sum_{w \in \Omega} \left| \Pr_P(w) - \Pr_Q(w) \right|.$$

We say that P is *ϵ-close* to Q, denoted by $P \overset{\epsilon}{\sim} Q$, if $|P - Q| \leq \epsilon$. We denote the fact that P and Q are identically distributed by $P \sim Q$. The following Lemma is trivial:

Lemma 4.1. *Let P, V be distributions on a set Ω. Suppose, $P = \delta \cdot R + (1 - \delta) \cdot V$ for two distributions R and V and $0 < \delta < 1$. Then $P \overset{\delta}{\sim} V$.*

We use *min-entropy* to measure the amount of randomness in a given distribution:

Definition 4.6 (min-entropy). *Let X be a distribution over a finite set Γ. The min-entropy of X is defined as*

$$H_\infty(X) \triangleq \min_{x \in supp(X)} \log \left(\frac{1}{\Pr[X = x]} \right)$$

Another useful measure of entropy is *collision probability.*

Definition 4.7 (collision probability). *Let X be a distribution over a finite set Γ. The* collision probability *of X is defined as*

$$cp(X) \triangleq \sum_{x \in supp(X)} \Pr[X = x]^2 = \Pr_{x_1, x_2 \leftarrow X}[x_1 = x_2].$$

The following lemma gives us a quantitative translation between the two quantities of min-entropy and collision probability.

Lemma 4.2 (Lemma 3.6 in [3]). *Let X be a distribution over a finite set Γ. Suppose that $cp(X) \leq \frac{1}{a \cdot b}$. Then X is $\frac{1}{\sqrt{a}}$-close to a distribution with min-entropy at least* $\log(b)$.

4.2.2 Polynomials over Finite Fields

We review some basic notions regarding polynomials defined over finite fields. Readers not familiar with the subject can find a more comprehensive treatment in [39]. For a field $\mathbb{F}$ we denote by $\mathbb{F}[t_1, \ldots, t_k]$ the ring of polynomials in k-variables $t_1, \ldots, t_k$ with coefficients in $\mathbb{F}$. We denote by $\mathbb{F}(t_1, \ldots, t_k)$ the field of rational functions in variables $t_1, \ldots, t_k$. We denote by $\deg(f)$ the total degree of f and by $\deg_{t_j}(f)$ the degree of f as a polynomial in t_j. We write $f \equiv 0$ or $f(t) \equiv 0$ if f is the zero polynomial (all coefficients of f are zero). Note that over the finite field $\mathbb{F}$ of prime cardinality p, the polynomial $f(t) = t^p - t$ is *not* the zero polynomial even though $f(a) = 0$ for all $a \in \mathbb{F}$.

We say that the polynomials $f_1, \ldots, f_m \in \mathbb{F}[t_1, \ldots, t_k]$ are *algebraically dependent* if there exists a nonzero polynomial $h \in \mathbb{F}[z_1, \ldots, z_m]$ such that $h(f_1(t), \ldots, f_m(t)) \equiv 0$. We sometimes refer to this polynomial h as the *annihilating polynomial* of $f_1, \ldots, f_m$. We say that $f_1, \ldots, f_m$ are *algebraically independent* if such a polynomial h does not exist.

For a polynomial $f \in \mathbb{F}[t_1, \ldots, t_k]$ we denote by $\frac{\partial f}{\partial t_j} \in \mathbb{F}[t_1, \ldots, t_k]$ the formal partial derivative of f with respect to the variable t_j. When using derivatives over a finite field we should be careful of 'strange' behavior of the derivative. For example, the derivative of t^p over a field of characteristic p is equal to zero. This is "strange" since t^p is not a constant function (in fact, it is a permutation). The following claim, which we use implicitly in many of our proofs, describes the exact conditions under which this "strange" behavior happens.

Claim 4.7.1. *Let $\mathbb{F}$ be a field of characteristic p and let $f \in \mathbb{F}[t_1, \ldots, t_k]$ and $j \in [k]$ be such that $\frac{\partial f}{\partial t_j} \equiv 0$. Then all degrees of t_j appearing in f are multiples of p. In particular, if $\deg_{t_j}(f) < p$, then $\frac{\partial f}{\partial t_j} \equiv 0$ iff $\deg_{t_j}(f) = 0$.*

For a vector of polynomials $\bar{f} = (f_1, \ldots, f_m) \in (\mathbb{F}[t_1, \ldots, t_k])^m$ we can define the *partial derivative matrix* of $\bar{f}$ as

$$\frac{\partial \bar{f}}{\partial t} \triangleq \begin{pmatrix} \frac{\partial f_1}{\partial t_1} & \cdots & \frac{\partial f_1}{\partial t_k} \\ \vdots & \ddots & \vdots \\ \frac{\partial f_m}{\partial t_1} & \cdots & \frac{\partial f_m}{\partial t_k} \end{pmatrix}$$

We denote by $\mathrm{rank}(\bar{f})$ the rank over $\mathbb{F}(t_1, \ldots, t_k)$ of the matrix $\frac{\partial \bar{f}}{\partial t}$.

Another useful property of polynomials, which we will use often, is the bound on the number of roots they can have. This generalization of the fundamental theorem of algebra is due to Schwartz and Zippel [60, 76].

Lemma 4.3 (Schwartz-Zippel). *Let $\mathbb{F}$ be a field and let $f \in \mathbb{F}[t_1, \ldots, t_k]$ be a nonzero polynomial with $\deg(f) \leq d$. Then, for any finite subset $S \subset \mathbb{F}$ we have*

$$\left|\left\{c \in S^k : f(c) = 0\right\}\right| \leq d \cdot |S|^{k-1}.$$

A simple corollary of the Schartz-Zippel Lemma is the following claim:

Claim 4.7.2. *Let $\mathbb{F}$ be a finite field and let $f \in \mathbb{F}[t_1, \ldots, t_k]$ be a polynomial of total degree at most d. Fix any $1 < i \leq k$. For $c = (c_i, \ldots, c_k) \in \mathbb{F}^{k-i+1}$ define*

$$f_c(t_1, \ldots, t_{i-1}) \triangleq f(t_1, \ldots, t_{i-1}, c_i, \ldots, c_k).$$

Then

$$\Pr_{c \leftarrow \mathbb{F}^{k-i+1}}(f_c \equiv 0) \leq \frac{d}{|\mathbb{F}|}.$$

4.2.3 The Number of Solutions to a System of Polynomial Equations

We will use a version of Bezout's Theorem proved by Wooley [73]. This theorem, mentioned informally in the Introduction, will give us a connection between algebraic rank and min-entropy. We note that the formulation of Wooley's theorem stated here is weaker than the original formulation appearing in [73] (the original form of the theorem speaks of congruences modulo p^s for any s).

Theorem 4.1 (rephrased from Theorem 1 in [73]). *Let $\mathbb{F}$ be a field of prime cardinality p. Let k and d be integers. Let $x = (x_1, \ldots, x_k) \in \mathcal{M}(\mathbb{F}^k \to \mathbb{F}^k, d)$ be such that rank$(x) = k$ and let $J(t) \triangleq \det\left(\frac{\partial x}{\partial t}\right)(t)$. For $a \in \mathbb{F}^k$ let*

$$N_a \triangleq \left|\left\{c \in \mathbb{F}^k\_: \_x(c) = a\_and\_J(c) \neq 0\right\}\right|.$$

Then for every $a \in \mathbb{F}^k$, $N_a \leq d^k$.

We can interpret this theorem as saying that a distribution X sampled by a non-degenerate mapping $x \in \mathcal{M}(\mathbb{F}^k \to \mathbb{F}^k, d)$ is close to a distribution with high min-entropy, where the closeness is related to the number of zeros of the determinant of $\frac{\partial x}{\partial t}$. Since this determinant is a nonzero low-degree polynomial, we get that the distance from the high min-entropy distribution is small. This is stated more precisely by the following corollary, which also extends our view to mappings in $\mathcal{M}(\mathbb{F}^k \to \mathbb{F}^n, d)$ for $k \leq n$.

Corollary 4.3. *Let $\mathbb{F}$ be a field of prime cardinality. Let $k \leq n$ and d be integers such that $|\mathbb{F}| > 2dk$. Let X be an (n, k, d)-polynomial source over $\mathbb{F}$. Then X is ϵ-close to a distribution with min-entropy at least $k \cdot \log\left(\frac{|\mathbb{F}|}{2d}\right)$, where $\epsilon = \frac{d \cdot k}{|\mathbb{F}|}$.*

Proof. X is the distribution $x(U_k)$ for a non-degenerate mapping $x \in \mathcal{M}(\mathbb{F}^k \to \mathbb{F}^n, d)$. Since x has rank k, the matrix $\frac{\partial x}{\partial t}$ has a nonsingular square submatrix. W.l.o.g assume that this matrix is composed of the first k rows of $\frac{\partial x}{\partial t}$. Let us also denote the determinant of this submatrix by $J(t)$.

Denote by C the event that $J(t) = 0$ and let $\delta = \Pr_{t \leftarrow \mathbb{F}^k}(C)$. Write X as a convex combination of conditional distributions as follows:

$$X = \delta \cdot (X|C) + (1 - \delta) \cdot (X|\neg C).$$

Note that, since $J(t)$ is a nonzero polynomial of degree at most $d \cdot k$, we have that $\delta \leq \frac{d \cdot k}{|\mathbb{F}|}$.

We claim that the distribution $(X|\neg C)$ has min-entropy at least $k \cdot \log(|\mathbb{F}|/2d)$: For any $a \in \mathbb{F}^n$, using Theorem 4.1

$$\Pr(X = a|\neg C) = \frac{\Pr(X = a \wedge \neg C)}{1 - \delta} \leq \frac{d^k}{|\mathbb{F}|^k \cdot (1 - \delta)}$$

$$\leq \frac{d^k}{|\mathbb{F}|^k \cdot (1 - dk/|\mathbb{F}|)} \leq \frac{2d^k}{|\mathbb{F}|^k} = \left(\frac{2d}{|\mathbb{F}|}\right)^k,$$

where we used the assumption about $|\mathbb{F}|$ in the last inequality. Thus, $(X|\neg C)$ has min-entropy at least $k \cdot \log(|\mathbb{F}|/2d)$, and using Lemma 4.1 we are done. □

4.3 Algebraic Independence and Rank

In [21] it is shown that, over the complex numbers, the two notions of rank and algebraic independence are equivalent. That is, the polynomials $x_1, \ldots, x_r \in \mathbb{F}[t_1, \ldots, t_k]$ are algebraically independent iff the matrix $\frac{\partial x}{\partial t}$ has maximal rank. In this section we prove two theorems showing that this connection is also valid over finite fields, provided the characteristic of the field is sufficiently large. We start by showing that maximal rank implies algebraic independence. This direction does not require the field characteristic to be large.

Theorem 4.2. *Let $\mathbb{F}$ be a field of characteristic p. Let $x = (x_1, \ldots, x_r) \in \mathcal{M}(\mathbb{F}^k \to \mathbb{F}^r, d)$ for some d, where $r \leq k$. If x has rank r then $x_1, \ldots, x_r$ are algebraically independent.*

Proof. Assume for contradiction that $x_1, \ldots, x_r$ are algebraically dependent. Let $g(z_1, \ldots, z_r)$ be a nonzero polynomial of minimal degree such that $g(x_1(t), \ldots, x_r(t)) \equiv 0$. Denote $g_i = \frac{\partial g}{\partial z_i}$.

Claim 4.7.3. *For some $1 \leq i \leq k$, g_i is nonzero.*

Proof. Fix some $1 \leq i \leq k$. Assume that $g_i \equiv 0$. Then, by Claim 4.7.1, all nonzero powers of z_i in g are multiples of p. Assume for contradiction that for all i, $g_i \equiv 0$. Then $g = h^p$ for some $h(z_1, \ldots, z_r)$, and

$$(h(x_1(t), \ldots, x_r(t)))^p \equiv 0 \Rightarrow h(x_1(t), \ldots, x_r(t)) \equiv 0,$$

and this is a contradiction to the minimality of g. □

We will go on to show that the derivatives of g form a nontrivial vector which is orthogonal to all the columns of $\frac{\partial x}{\partial t}$, contradicting our assumption that $\frac{\partial x}{\partial t}$ has maximal rank. Using the above claim, fix an i such that g_i is nonzero. By the minimality of the degree of g we know that $g_i(x_1(t), \ldots, x_r(t))$

is nonzero as a polynomial in t (the degree of the derivative is always smaller than that of the original polynomial). Define $\overline{g}(t) \triangleq g(x_1(t), \ldots x_r(t))$. Note that $\overline{g}(t) \equiv 0$. Using the chain rule, for $1 \leq j \leq k$ we have

$$0 = \frac{\partial \overline{g}}{\partial t_j} = \sum_{l=1}^{r} g_l(x(t)) \cdot \frac{\partial x_l}{\partial t_j}.$$

Note that the rightmost expression is the inner product of the nonzero vector

$$u = (g_1(x(t)), \ldots, g_r(x(t)))$$

and the jth column of the matrix $\frac{\partial x}{\partial t}$. Thus, we have

$$u \cdot \frac{\partial x}{\partial t} = 0$$

for $u \neq 0$, and so the rank of $\frac{\partial x}{\partial t}$ is at most $r - 1$, a contradiction. □

We now turn to prove the other direction, which states that algebraic independence implies maximal rank. In order to prove this direction we require the field characteristic to be larger than $(k+1)d^k$ where k is the number of variables and d is the total degree of the polynomials. This requirement stems from the degree of the annihilating polynomial we find in the proof. Our proof is based on the same ideas appearing in [21, 43, 73]. We are not aware how tight the degree bound we get in the proof is. Another approach is to use Gröbner Bases, which often leads to double exponential degrees.

Theorem 4.3. *Let $\mathbb{F}$ be a field of characteristic p. Let d, k and n be integers such that $p > D$, where $D = (k+1) \cdot d^k$. Let $x \in \mathcal{M}(\mathbb{F}^k \to \mathbb{F}^n, d)$ have rank smaller than n. Then, there exists a nonzero polynomial $h \in \mathbb{F}[z_1, \ldots, z_n]$ of total degree at most D such that*

$$h(x_1(t), \ldots, x_n(t)) \equiv 0.$$

Proof. Fix any d and k. We first prove the theorem for $n \geq k+1$. Assume w.l.g. that $n = k+1$ (if $n > k+1$ we can use this case to find an h that uses only the first $k+1$ variables). In this case, the coefficients of the required h can be found by showing that a certain system of linear equations has more degrees of freedom than constraints. More precisely, we want a nonzero polynomial h of degree at most D such that $\overline{h}(t) \triangleq h(x_1(t), \ldots, x_n(t)) \equiv 0$. The number of constraints is the number of coefficients of $\overline{h}$. Since $deg(\overline{h}) \leq d \cdot D$, this is at most $\binom{d \cdot D+k}{k}$. The number of variables is the number of coefficients of h, which is $\binom{D+n}{n} = \binom{D+k+1}{k+1}$. We show that the number of variables is larger than the number of constraints:

$$\binom{D+k+1}{k+1} \Big/ \binom{d \cdot D + k}{k} = \frac{(D+k+1)!}{D!(k+1)!} \cdot \frac{k!(d \cdot D)!}{(d \cdot D + k)!}$$

$$= \frac{(D+1)\cdots(D+k+1)}{(k+1)\cdot(d\cdot D+1)\cdots(d\cdot D+k)} \geq \left(\frac{D}{d\cdot D}\right)^k \cdot \frac{D+k+1}{k+1}$$
$$= \frac{D+k+1}{d^k\cdot(k+1)} > 1.$$

We now prove the claim for $n \leq k$ by backwards induction on n. We assume the claim for $n+1$ and prove it for n. Assume for contradiction that there is no nonzero polynomial $h(z_1,\ldots,z_n)$ of degree at most D such that $h(x_1(t),\ldots,x_n(t)) \equiv 0$. Using the induction hypothesis, for each $1 \leq i \leq k$ we have a nonzero polynomial $h_i(z_1,\ldots,z_n,w)$ of degree at most D with

$$h_i(x_1(t),\ldots,x_n(t),t_i) \equiv 0. \tag{4.1}$$

We will go on to show that the partial derivatives of the polynomials h_i form a matrix which is the 'inverse' of $\frac{\partial x}{\partial t}$, contradicting our assumption about the rank of $\frac{\partial x}{\partial t}$. W.l.o.g assume that h_i is a minimal degree polynomial satisfying (4.1). For $1 \leq j \leq n$ denote $h_{i,j} = \frac{\partial h_i}{\partial z_j}$ and let $h_{i,0} = \frac{\partial h_i}{\partial w}$. By our contradiction assumption, h_i must contain nonzero powers of w, and since $deg(h_i) < p$ this implies that $h_{i,0}$ is nonzero. By the minimality of the degree of h_i, we have that $h_{i,0}(x_1(t),\ldots,x_n(t),t_i)$ is a *nonzero* polynomial in t. Taking the derivative of (4.1) for each $1 \leq l \leq k$, we have

$$0 = \sum_{j=1}^{n} h_{i,j}\cdot\frac{\partial x_j}{\partial t_l} + \delta_{i,l}\cdot h_{i,0}.$$

Since we can divide by the nonzero $h_{i,0}$ we get

$$\frac{-1}{h_{i,0}}\sum_{j=1}^{n} h_{i,j}\cdot\frac{\partial x_j}{\partial t_l} = \delta_{i,l}$$

for every $1 \leq i \leq k$ and $1 \leq l \leq k$. Therefore, we have $H\cdot\frac{\partial x}{\partial t} = I$, where H is the $k\times n$ matrix with $H_{i,j} = \frac{-h_{i,j}}{h_{i,0}}$, contradicting the assumption that $\frac{\partial x}{\partial t}$ has rank smaller than n. □

4.4 An Explicit Rank Extractor

In this section we describe our construction of a rank extractor and prove Theorem 4.1.

Construction 1. *Let $k \leq n$ and d be integers. Let $s_2 = dk+1$ and $s_1 = (2dn+1)\cdot s_2$. Let $l_{ij} = i\cdot(s_1 + j\cdot s_2)$. Define for each $1 \leq i \leq k$*

$$y_i(x) = y_i(x_1,\ldots,x_n) \triangleq \sum_{j=1}^{n} \frac{1}{l_{ij}+1}\cdot x_j^{l_{ij}+1}.$$

Let $y = (y_1, \ldots, y_k)$ be the output of the construction. Notice that $y(x)$ is defined in such a way that the partial derivative $\frac{\partial y_i}{\partial x_j}$ is exactly $x_j^{l_{ij}}$.

We prove the following theorem, which directly implies Theorem 4.1.

Theorem 4.4. *Let $\mathbb{F}$ be a field of characteristic zero or of characteristic larger than $d' = 8k^2d^3n$. Let $x \in \mathcal{M}(\mathbb{F}^k \to \mathbb{F}^n, d)$ be of rank k. Let $y : \mathbb{F}^n \to \mathbb{F}^k$ be as in Construction 1. Then the composition $(y \circ x)(t)$ is in $\mathcal{M}(\mathbb{F}^k \to \mathbb{F}^k, d')$ and has rank k.*

4.4.1 Preliminaries for the Proof of Theorem 4.4

Sums of powers of polynomials

The following lemma shows how to pick integers $c_1, \ldots, c_n$ in such a way that for any set of n polynomials $x_1(t), \ldots, x_n(t)$ of bounded degree, the polynomials $x_1(t)^{c_1}, \ldots, x_n(t)^{c_n}$ will have degrees that are different by at least some fixed number.

Lemma 4.4. *Let $x_1(t), \ldots, x_n(t)$ be k-variate non-constant polynomials over some field $\mathbb{F}$. Denote by $d_i > 0$ the degree of the polynomial x_i. Let $d \geq \max_i\{d_i\}$. Let A and B be two positive integers such that $A \geq (2dn+1) \cdot B$ and let $c_i \triangleq A + Bi$ for $i \in [n]$. Then, for every $1 \leq i < j \leq n$, we have*

$$|\deg(x_i(t)^{c_i}) - \deg(x_j(t)^{c_j})| = |d_i \cdot c_i - d_j \cdot c_j| \geq B.$$

Proof. Let $1 \leq i < j \leq n$. First, suppose that $d_i = d_j$. In this case we have

$$d_j \cdot c_j - d_i \cdot c_i = d_j(A + Bj) - d_i(A + Bi) = d_j \cdot B \cdot (j - i) \geq B.$$

Next, suppose $d_j \neq d_i$. In this case we have

$$\begin{aligned} |d_j \cdot c_j - d_i \cdot c_i| &= |d_j(A + Bj) - d_i(A + Bi)| \\ &= |(d_j - d_i)A + d_jBj - d_iBi| \\ &\geq |d_j - d_i|A - |d_jBj| - |d_iBi| \\ &\geq A - 2dnB \geq B. \end{aligned}$$

□

The Cauchy-Binet formula

The Cauchy-Binet formula gives the determinant of the product of a $k \times n$ matrix with an $n \times k$ matrix (for $k \leq n$). Let $k \leq n$. Let A be a $k \times n$ matrix and B be an $n \times k$ matrix. For a set $I \subset [n]$ of size k we denote by A_I the $k \times k$ submatrix of A composed of the columns of A whose indices appear in I. Similarly, we denote by B_I the submatrix of B composed of the rows of B whose indices are in I. The proof of the following formula can be found in [28].

Lemma 4.5 (Cauchy-Binet). *Let $k \le n$. Let A be a $k \times n$ matrix and B an $n \times k$ matrix over a field $\mathbb{F}$. Using the above notations we have*

$$\det(A \cdot B) = \sum_{\substack{I \subset [n] \\ |I|=k}} \det(A_I) \cdot \det(B_I).$$

4.4.2 Proof of Theorem 4.4

Let $k \le n$, d be integers. Let $\mathbb{F}$ be a field of characteristic zero or of characteristic larger than $d' = 8k^2d^3n$. Let $x = (x_1, \ldots, x_n) \in \mathcal{M}(\mathbb{F}^k \to \mathbb{F}^n, d)$ be such that $\operatorname{rank}(x) = k$. Let $y : \mathbb{F}^n \to \mathbb{F}^k$ be defined as in Construction 1, that is,

$$y_i(x) = y_i(x_1, \ldots, x_n) \triangleq \sum_{j=1}^{n} \frac{1}{l_{ij}+1} \cdot x_j^{l_{ij}+1}, \tag{4.2}$$

where

$$l_{ij} = i \cdot (s_1 + j \cdot s_2)$$

$$s_1 = (2dn+1) \cdot s_2, \; s_2 = dk + 1.$$

It is easy to verify that the degree of the mapping y is bounded by $8k^2d^2n$. Therefore, the degree of the composition $(y \circ x)(t)$ is bounded by $d' = 8k^2d^3n$. Therefore, since the characteristic of $\mathbb{F}$ is larger than d' (or is zero), for the rest of the proof we don't need to worry about non-constant polynomials becoming zero after we take their derivative (see Claim 4.7.1).

Our goal is to show that the composition $y \circ x$ has rank k. In order to prove this we need to show that the determinant of the partial derivatives matrix of the composition is nonzero. Write $y(t)$ to denote $y(x(t))$ and let $\frac{\partial y}{\partial t}$ denote the $k \times k$ partial derivative matrix of the mapping $y(t)$. Using the chain rule we have that

$$\frac{\partial y}{\partial t} = \frac{\partial y}{\partial x} \cdot \frac{\partial x}{\partial t},$$

where $\frac{\partial y}{\partial x}$ is a $k \times n$ matrix and $\frac{\partial x}{\partial t}$ is an $n \times k$ matrix. All the elements in these two matrices are polynomials in t, since we evaluate $\frac{\partial y}{\partial x}$ at $x = x(t)$.

Consider the element at position (i, j) in the matrix $\frac{\partial y}{\partial x}$. Taking the derivative of (4.2) with respect to x_j we get that

$$\frac{\partial y_i}{\partial x_j} = x_j(t)^{l_{ij}} = x_j(t)^{i \cdot (s_1 + j s_2)}.$$

The Vandermonde structure of $\frac{\partial y}{\partial x}$ becomes more apparent by letting $r_j(t) \triangleq x_j(t)^{s_1 + j s_2}$. We now have that the (i, j)th element of $\frac{\partial y}{\partial x}$ is $r_j(t)^i$. That is,

$$\frac{\partial y}{\partial x} = \begin{pmatrix} r_1(t) & r_2(t) & \cdots & \cdots & r_n(t) \\ r_1(t)^2 & r_2(t)^2 & \ddots & & r_n(t)^2 \\ \vdots & \vdots & & \ddots & \vdots \\ r_1(t)^k & r_2(t)^k & \cdots & \cdots & r_n(t)^k \end{pmatrix}$$

To facilitate presentation, let $R \triangleq \frac{\partial y}{\partial x}$ and $D \triangleq \frac{\partial x}{\partial t}$. We can also assume w.l.o.g that

$$\deg(r_1(t)) \leq \ldots \leq \deg(r_n(t)) \tag{4.3}$$

(we let $\deg(0) = 0$) since applying the same permutation on the rows of R and on the columns of D will not change the determinant of $R \cdot D$. Now, from Lemma 4.5 (Cauchy-Binet) and using the notations of Section 4.4.1, we have that

$$\det\left(\frac{\partial y}{\partial t}\right) = \det(R \cdot D) = \sum_{\substack{I \subset [n] \\ |I|=k}} \det(R_I) \cdot \det(D_I.) \tag{4.4}$$

Notice that if $r_i(t)$ is constant, then $x_i(t)$ is also constant and so the ith row of the matrix D is zero. Therefore, $\det(D_I) = 0$ for every I that contains an index i such that $r_i(t)$ is constant. In view of (4.4) and this last observation, we can assume w.l.o.g that for all $i \in [n]$, $r_i(t)$ is non-constant. (Notice that since D has maximal rank, we have at least k indices in $[n]$ for which $x_i(t)$ is non-constant, and so the condition $n \geq k$ is maintained).

The next three claims will show that there exist a unique set I in the above sum for which the degree of $\det(R_I) \cdot \det(D_I)$ is maximal. This will conclude the proof, since then we will have that $\det\left(\frac{\partial y}{\partial t}\right)$ is nonzero, as required.

We start with a simple claim showing that the degrees of the polynomials $r_i(t)$ have large gaps between them.

Claim 4.7.4. *Let $r_1(t), \ldots, r_n(t)$ be the polynomials defined above. Then for every $i \in [n-1]$ we have*

$$\deg(r_{i+1}(t)) > \deg(r_i(t)) + dk.$$

Proof. Recall that $r_i(t) = x_i(t)^{s_1 + j \cdot s_2}$ and that $s_1 \geq (2dn+1) \cdot s_2$. Using Lemma 4.4 we get that

$$|\deg(r_{i+1}(t)) - \deg(r_i(t))| \geq s_2 > dk.$$

Using (4.3) the claim follows. □

Let $I \subset [n]$ be such that $|I| = k$. We let by

$$d_I \triangleq \deg\left(\det(R_I)\right)$$

The next claim gives a convenient formula for d_I.

Claim 4.7.5. *Let $I \subset [n]$, $I = \{i_1 < \ldots < i_k\}$. Then*

$$d_I = \deg(R_I) = \sum_{j=1}^{k} j \cdot \deg\left(r_{i_j}(t)\right)$$

Proof. Using the Vandermonde structure of the matrix R_I we get that

$$\det(R_I) = \prod_{j=1}^{k} r_{i_j}(t) \prod_{1 \leq j_1 < j_2 \leq k} \left(r_{i_{j_1}}(t) - r_{i_{j_2}}(t)\right).$$

In view of (4.3), the degree of the highest monomial in $\det(R_I)$ is obtained my multiplying k copies of $r_{i_k}(t)$ with $k-1$ copies of $r_{i_{k-1}}(t)$, and so on. This will give a monomial with degree $\sum_{j=1}^{k} j \cdot \deg(r_j(t))$. □

Define

$$\Gamma \triangleq \{I \subset [n] \mid |I| = k\,,\, \det(D_I) \neq 0\}\,.$$

The next and final claim shows that there exists a *unique* $I \in \Gamma$ with maximal d_I. The proof uses standard techniques from matroid theory.

Claim 4.7.6. *Let $d_{\max} \triangleq \max_{I \in \Gamma}\{d_I\}$. Then there exists a unique $I^* \in \Gamma$ such that $d_{I^*} = d_{\max}$. Moreover, for every $I \neq I^*$ we have that $d_I < d_{I^*} - dk$.*

Proof. Let $v_1, \ldots, v_n$ denote the rows of D. We can treat $v_1, \ldots, v_n$ as vectors in a k-dimensional vector space over the field of rational functions in variables $t_1, \ldots, t_k$.

We are going to construct the set I^* using the following greedy algorithm: Start with $I^* = \emptyset$ and at each step add to I^* the largest $i \in [n]$ for which the set $\{\, v_j \mid j \in I^* \cup \{i\}\,\}$ is linearly independent. Since we assumed that D has maximal rank, this process will end after precisely k steps, yielding a set I^* of size k and such that $\det(D_{I^*}) \neq 0$. Denote by $I^* = \{i_1^* < \ldots < i_k^*\}$.

Observing the formula for d_I given by Claim 4.7.5 and recalling that the degrees of the polynomials r_i are strictly increasing, we see that the greedy construction of I^* ensures that $d_{I^*} = d_{\max}$. Assume in contradiction that there exists a set $I' \neq I$ in Γ such that $d_{I'} = d_{\max}$ and let $I' = \{i'_1 < \ldots < i'_k\}$. From the monotonicity of $\deg(r_i(t))$ it follows that there must be an index $j \in [k]$ such that $i'_j > i^*_j$ (otherwise we would have $d_{I'} < d_{I^*}$). Let $j' \in [k]$ be the largest index such that $i'_{j'} > i^*_{j'}$. Since $I' \in \Gamma$ we have that the set $\left\{v_{i'_{j'}}, v_{i'_{j'+1}}, \ldots, v_{i'_k}\right\}$ is linearly independent. Therefore there must be an index $0 \leq \alpha \leq k - j'$ such that the vector $v_{i'_{j'+\alpha}}$ is not spanned by the set of vectors $\left\{v_{i^*_{j'+1}}, v_{i^*_{j'+2}}, \ldots, v_{i^*_k}\right\}$. This contradicts the greedy construction of I^* since, by construction, all the vectors $v_{i^*_{j'}+1}, v_{i^*_{j'}+2}, \ldots, v_n$ are spanned by $\left\{v_{i^*_{j'+1}}, v_{i^*_{j'+2}}, \ldots, v_{i^*_k}\right\}$.

To prove the "moreover" part of the claim we use Claim 4.7.4. Let $I = \{i_1 < \ldots < i_k\}$ be such that $I \neq I^*$ and $I \in \Gamma$. Using the same logic as above we can deduce that for all $j \in [k]$, $i_j \leq i_j^*$ and that for some $j' \in [k]$, $i_{j'} < i_{j'}^*$. Plugging this information into the formula for d_I we get that

$$\begin{aligned} d_{I^*} - d_{I'} &= \sum_{j=1}^{k} j \cdot \left(\deg\left(r_{i_j^*}(t)\right) - \deg\left(r_{i_j}(t)\right)\right) \\ &\geq \deg\left(r_{i_{j'}^*}(t)\right) - \deg\left(r_{i_{j'}}(t)\right) \\ &> dk, \end{aligned}$$

where the last inequality follows from Claim 4.7.4. □

We can now use Claim 4.7.6 to show that the sum in (4.4) is not zero. Let $I^* \in \Gamma$ be the set with unique maximal d_{I^*} given by Claim 4.7.6. Rewrite (4.4) in the following form

$$\begin{aligned} \det(R \cdot D) &= \sum_{\substack{I \subset [n] \\ |I|=k}} \det(R_I) \cdot \det(D_I) \\ &= \sum_{I \in \Gamma} \det(R_I) \cdot \det(D_I) \\ &= \det(R_{I^*}) \cdot \det(D_{I^*}) + \sum_{\substack{I \in \Gamma \\ I \neq I^*}} \det(R_I) \cdot \det(D_I). \end{aligned} \tag{4.5}$$

The degree of the first summand in (4.5) is at least

$$\deg\left(\det(R_{I^*}) \cdot \det(D_{I^*})\right) = d_{I^*} + \deg\left(\det(D_{I^*})\right) \geq d_{I^*}.$$

Using Claim 4.7.6 we can upper bound the degrees of the other summands in (4.5). That is, for all $I \in \Gamma$ different from I^* we have

$$\deg\left(\det(R_I) \cdot \det(D_I)\right) = d_I + \deg\left(\det(D_I)\right) \leq d_I + dk < d_{I^*}$$

(we use the fact that all the entries of D are polynomials of degree at most d). Therefore, the sum in (4.5) cannot be zero. This concludes the proof of Theorem 4.4. □

4.5 Extractors for Polynomial Sources

In this section we describe our construction of an extractor for full-rank polynomial sources and prove Theorem 4.2. As was mentioned in the introduction, this construction, together with the rank extractor constructed in previous sections, will give an extractor for polynomial sources of any rank. In order to describe our construction we require some additional notations. Let $\mathbb{F}$ be a field of prime cardinality p. For an integer $M \leq p$, we denote by $\mathrm{mod}_M : \mathbb{F} \to \{0, \ldots, M-1\}$ the modulo-M function. For a vector $x \in \mathbb{F}^n$ we apply the function $\mathrm{mod}_M(x)$ coordinate-wise. The following theorem directly implies Theorem 4.2.

Theorem 4.5. *There exist absolute constants $C > 0$ and $c > 0$ such that the following holds. Let k, d be integers and let $\mathbb{F}$ be a field of prime cardinality $p > d^{Ck}$. Let $m > 0$ be an integer such that $m < c \cdot log(p)$, let $M = 2^m$ and define the function $E : \mathbb{F}^k \to \{0,1\}^{km}$ as $E(y) \triangleq mod_M(y)$. Then for every (k, k, d)-polynomial source Y over $\mathbb{F}$, the distribution $E(Y)$ is ϵ-close to uniform with $\epsilon = p^{-\Omega(1)}$.*

Notice that the construction of the extractor is very simple— taking a module in each coordinate. Proving that this is an extractor is much more complicated. The main tool in the proof of Theorem 4.5 will be a theorem of Bombieri [8] giving an exponential sum estimate for low degree polynomials defined over curves (one-dimensional varieties). We refer the reader to Appendix B for a discussion of the basic notions of algebraic geometry used in the proof.

4.5.1 Preliminaries for the Proof of Theorem 4.5

Block distributions

Our proof will rely on the following standard lemmas concerning block distributions.

Lemma 4.6. *Let A be some finite set and let $X = (X_1, \ldots, X_k)$ be a distribution on A^k. Let $0 < \epsilon < 1$ and suppose that X_1 is ϵ-close to uniform. Suppose also that for each $2 \leq i \leq k$ there exists a set $S_i \subset A^{i-1}$ such that*

1. $\Pr[(X_1, \ldots, X_{i-1}) \in S_i] \geq 1 - \epsilon$ *and*
2. *For each $s \in S_i$, the conditional distribution $(X_i | (X_1, \ldots, X_{i-1}) = s)$ is ϵ-close to uniform.*

Then X is $O(k \cdot \epsilon)$-close to uniform.

Proof. We will prove the lemma for $k = 2$ (the general case will follow by a straightforward induction). Let $T \subset A^2$ be some non empty set. It suffices to show that $\left|\Pr[(X_1, X_2) \in T] - |T|/|A|^2\right| \le O(\epsilon)$. For each $a \in A$ let $T_a = T \cap (\{a\} \times A)$. Let $S = S_2 \subset A$ be the set from the lemma. We have that

$$\begin{aligned}
\Pr[(X_1, X_2) \in T] &= \sum_{a \in A} \Pr[X_1 = a] \cdot \Pr[X_2 \in T_a | X_1 = a] \\
&\le \epsilon + \sum_{a \in S} \Pr[X_1 = a] \cdot \Pr[X_2 \in T_a | X_1 = a] \\
&\le 2\epsilon + \sum_{a \in S} \Pr[X_1 = a] \cdot \frac{|T_a|}{|A|} \\
&\le 3\epsilon + \sum_{a \in A} \frac{|T_a|}{|A|^2} = 3\epsilon + \frac{|T|}{|A|^2}.
\end{aligned}$$

Similarly, we can show an inequality in the opposite direction and so we conclude that (X_1, X_2) is 3ϵ-close to uniform. □

For our proof we require a modified version of this last lemma. In the modified version we fix not only the prefix of the distribution, but rather all indices except the ith one. We recall our notation that for a vector $v = (v_1, \ldots, v_n)$ and for an index $i \in [n]$ we have $v^{(-i)} = (v_1, \ldots, v_{i-1}, v_{i+1}, \ldots, v_n)$. In some places we will define a new vector of length $n-1$ by writing $u = u^{(-i)} \in A^{n-1}$. This means that the indices of u go from 1 to n, skipping the ith index. That is, $u = (u_1, \ldots, u_{i-1}, u_{i+1}, \ldots, u_n) \in A^{n-1}$.

Lemma 4.7. *Let A be some finite set and let $X = (X_1, \ldots, X_k)$ be a distribution on A^k. Let $0 < \epsilon < 1$ and suppose that for each $1 \le i \le k$ there exists a set $S_i \subset A^{k-1}$ such that*

1. $\Pr[X^{(-i)} \in S_i] \ge 1 - \epsilon$ *and*
2. *For each $s^{(-i)} \in S_i$, the conditional distribution $(X_i | X^{(-i)} = s^{(-i)})$ is ϵ-close to uniform.*

Then X is $O(k \cdot \sqrt{\epsilon})$-close to uniform.

Proof. The lemma will follow by showing that X satisfies the conditions of Lemma 4.6 with ϵ replaced by $O(\sqrt{\epsilon})$. The first block X_1 (and indeed, every other block) is easily seen to be 2ϵ close to uniform by breaking it into a convex combination over all fixings of the other blocks, and throwing away those fixings not in S_1.

Now, let $i > 1$. For a prefix $(a_1, \ldots, a_{i-1}) \in A^{i-1}$ we define $P(a_1, \ldots, a_{i-1})$ to be the probability that $a^{(-i)} = (a_1, \ldots, a_{i-1}, a_{i+1}, \ldots, a_k)$ is in S_i when the additional elements $(a_{i+1}, \ldots, a_k)$ are chosen according the the distribution $(X_{i+1}, \ldots, X_k | X_1 = a_1, \ldots, X_{i-1} = a_{i-1})$. A simple averaging argument shows that the set $S'_i = \{(a_1, \ldots, a_{i-1}) \mid P(a_1, \ldots, a_{i-1}) \ge 1 - \sqrt{\epsilon}\}$ has probability at least $1 - \sqrt{\epsilon}$ in the distribution of $(X_1, \ldots, X_{i-1})$. We can thus, apply Lemma 4.6 with the sets S'_i and with ϵ replaced by $2\epsilon + \sqrt{\epsilon} = O(\sqrt{\epsilon})$. □

Distributions with small Fourier coefficients

The following lemma is an extension of the now folklore Vazirani XOR Lemma [29] and is used [10, 5] to extract randomness from distributions with bounded Fourier coefficients. What the lemma says is that if we have a distribution X with a bound of $p^{-\Omega(1)}$ on all of its Fourier coefficients then we can deterministically extract from X (using the modulo function) $\Omega(\log(p))$ bits that are $p^{-\Omega(1)}$-close to uniform. The following formulation of the lemma follows from the version proved in [51].

Lemma 4.8. *Let p be a prime number and let $0 < \alpha < 1$ be such that $\log(p) < p^{\alpha/2}$. Let X be a distribution on $\mathbb{F}$, the field of p elements. Suppose that for every nontrivial additive character $\chi : \mathbb{F} \to \mathbb{C}^*$ we have the bound $\mathbb{E}[\chi(X)] \leq p^{-\alpha}$. Let $m = \lfloor(\alpha/2)\cdot\log(p)\rfloor$, let $M = 2^m$ and let $Y = mod_M(X)$ be an m-bit random variable. Then Y is $p^{-\alpha/4}$-close to uniform.*

Intersections of hypersurfaces

Consider a system of $n-1$ polynomial equations in n variables. The next lemma gives a bound on the number of 'shifts' of the system for which the set of solutions has dimension larger than one (for the precise meaning of 'shift' see the lemma).

Lemma 4.9. *Let $\mathbb{F}$ be a finite field of size p and let $\bar{\mathbb{F}}$ denote its algebraic closure. Let $f_1, \ldots, f_{n-1} \in \mathbb{F}[x_1, \ldots, x_n]$ be polynomials of degree $\leq d$. For every $a = (a_1, \ldots, a_{n-1}) \in \mathbb{F}^{n-1}$ let $\hat{V}_a = \{x \in \bar{\mathbb{F}}^n \mid f_i(x) = a_i\,, i \in [n-1]\}$ and let $A = \{a \in \mathbb{F}^{n-1} \mid \hat{V}_a \neq \emptyset \text{ and } \dim(\hat{V}_a) \neq 1\}$. Then $|A| \leq nd^n p^{n-2}$.*

Proof. In order to bound $|A|$ we will describe an injective mapping from A to some small set. Fix some $a = (a_1, \ldots, a_{n-1}) \in A$. For $i \in [n-1]$ let $H_i = \{x \in \bar{\mathbb{F}}^n \mid f_i(x) = a_i\}$ be the hypersurface defined by the ith restriction and let $U_i = H_1 \cap \ldots \cap H_i$ so that $U_{n-1} = \hat{V}_a$. Using Lemma B.10 we see that if $\hat{V}_a$ is not empty and $\dim(\hat{V}_a) \neq 1$ then there must be some $2 \leq i \leq n-1$ such that H_i contains one of the irreducible components of U_{i-1}. Let i' be the smallest i satisfying this condition and let $0 < L \leq d^n$ be the index of the corresponding irreducible component of $U_{i'-1}$ (using some arbitrary ordering of the components of $U_{i'-1}$), where the bound of d^n on L follows from Lemma B.12. Observe that if we are given the set of values $\{a^{(-i')}, i', L\}$ we can determine $a_{i'}$ and so recover a. Therefore, there exists an injective mapping from A into the set $\mathbb{F}^{n-2} \times [n] \times [d^n]$. Therefore $|A| \leq nd^n \cdot p^{n-2}$. □

A theorem of Bombieri

The final ingredient we require for the proof of Theorem 4.5 is an exponential sum estimate due to Bombieri [8]. We quote here a weak version of Bombieri's theorem which is sufficient for our needs (see Appendix B for more details on this result).

Theorem 4.6 (Theorem 6 in [8]). *Let p be a prime and let $1 < d$ be an integer such that $d^n < p$. Let $\mathbb{F}$ be the field of p elements and let $\bar{\mathbb{F}}$ be its algebraic closure. Let $f_1, \ldots, f_{n-1} \in \mathbb{F}[x_1, \ldots, x_n]$ be $n-1$ polynomials of degree $\leq d$ such that the set $\hat{V} = \{x \in \bar{\mathbb{F}}^n | f_1(x) = \ldots = f_{n-1}(x) = 0\}$ is a curve. Let $g \in \mathbb{F}[x_1, \ldots, x_n]$ be a polynomial of degree $\leq d$ that is non-constant on at least one of the irreducible components of $\hat{V}$. Let $\hat{V} = \hat{V}_1 \cup \ldots \cup \hat{V}_L$ be the decomposition of $\hat{V}$ into irreducible components. Let $\hat{U}$ be the union of those irreducible components of $\hat{V}$ on which $g(x)$ is non-constant and let $U = \hat{U} \cap \mathbb{F}$. Let $\chi : \mathbb{F} \to \mathbb{C}^*$ be a nontrivial additive character of $\mathbb{F}$. Then*

$$\left| \sum_{x \in U} \chi(g(x)) \right| \leq 4d^{2n} \cdot p^{1/2}.$$

4.5.2 Proof of Theorem 4.5

Let $Y : \mathbb{F}^k \to \mathbb{F}^k$ be a (k, k, d)-polynomial source and let $f = (f_1, \ldots, f_k) \in \mathbb{F}[x_1, \ldots, x_k]$ be a vector of polynomials of degree at most d such that $Y(x) = f(x) = (f_1(x), \ldots, f_k(x))$. For $i \in [k]$ and $a = a^{(-i)} \in \mathbb{F}^{k-1}$, we let $V_a = \{x \in \mathbb{F}^k \,|\, f^{(-i)}(x) = a\}$ and also $\hat{V}_a = \{x \in \bar{\mathbb{F}}^k \,|\, f^{(-i)}(x) = a\}$, where $\bar{\mathbb{F}}$ denotes the algebraic closure of $\mathbb{F}$. For a nontrivial additive character $\chi : \mathbb{F} \to \mathbb{C}^*$, such that $V_a \neq \emptyset$ we define the exponential sum

$$\Upsilon_\chi(a) = \frac{1}{|V_a|} \sum_{x \in V_a} \chi(f_i(x)).$$

In view of Lemma 4.7 and 4.8 the theorem will follow from the following lemma.

Lemma 4.10. *Using the above notations, there exists $0 < \alpha < 1$ such that for every $i \in [k]$ there exists a set $S_i \subset \mathbb{F}^{k-1}$ such that*

1. *$f^{(-i)}(x)$ lands in S_i with probability at least $1 - p^{-\alpha}$ when x is chosen uniformly in $\mathbb{F}^k$.*
2. *For every $a = a^{(-i)} \in S_i$ and for every nontrivial χ, $|\Upsilon_\chi(a)| \leq p^{-\alpha}$.*

Before proving the lemma we proceed to show how it is used to complete the proof of Theorem 4.5. Let us denote by

$$Z_i = \text{mod}_M(f_i(x))$$

the random variable representing the ith block of $E(Y)$. Let $0 < \alpha < 1$ be the constant given by Lemma 4.10. Let $i \in [k]$ and let $S_i \subset \mathbb{F}^{k-1}$ be the set given by Lemma 4.10. We define the set $S'_i = \text{mod}_M(S_i)$ to be the image of S_i under the function $\text{mod}_M(\cdot)$. From part (1) of Lemma 4.10 we get that $Z^{(-i)}$ lands in S'_i with probability at least $1 - p^{-\Omega(1)}$. For $b =$

$b^{(-i)} \in [M]^{k-1}$ let $Z_i(b)$ be the random variable distributed according to the conditional distribution $(Z_i|Z^{(-i)} = b)$. The random variable $Z_i(b)$ is a convex combination of distributions $W_i(a) = (Z_i|f^{(-i)}(x) = a)$ taken over all $a = a^{(-i)}$ such that $\text{mod}_M(a) = b$. Since, by the definition of S'_i, these as are all in S_i, we can use part (2) of Lemma 4.10 together with Lemma 4.8 to get that each $W_i(a)$ in the convex combination of $Z_i(b)$ is $p^{-\Omega(1)}$-close to uniform. This, of course, holds then also for $Z_i(b)$. We finish the proof by observing that $Z = (Z_1, \ldots, Z_k)$ satisfies all the conditions of Lemma 4.7 with $\epsilon = p^{-\Omega(1)}$, and so we are done, since $O(k \cdot \sqrt{p^{-\Omega(1)}}) = p^{-\Omega(1)}$ when $p > d^{Ck}$ and C is sufficiently large.

Proof of Lemma 4.10

Let $i \in [k]$. We would like to distinguish between "good" and "bad" fixings of $f^{(-i)}(x)$. The "good" fixings will be those values $a = a^{(-i)} \in \mathbb{F}^{k-1}$ for which we can bound the exponential sum $\Upsilon_\chi(a)$. Before proving the Lemma formally let us describe briefly the intuition behind the proof. Each fixing $f^{(-i)}(x) = a^{(-i)}$ defines a variety V. We would like to apply Bombieri's theorem to bound the exponential sum of $f_i(x)$ over this variety. In order to do so we need to make sure that V is a curve and that $f_i(x)$ is not constant on "enough" of the components of the curve V (where the word "enough" takes into account the number of points in $\mathbb{F}$ in each component). The fact that most fixings satisfy the first condition, that V is a curve, will follow from a counting argument, based on a version of Bezout's theorem. The second condition will follow from Wooley's theorem (Theorem 4.1). Intuitively, Wooley's theorem tells us that the image of f is close to having high min-entropy. Clearly, this should allow us to bound the size of those components on which $f_i(x)$ is constant (for "most" fixings of $f^{(-i)}(x)$).

In order to be able to define these "good" fixings of $f^{(-i)}(x)$ we need to consider the singular points of the mapping $f(x)$, namely the zeros of its Jacobian. Let $J(x) = \det\left(\frac{\partial f}{\partial x}\right)$ be the determinant of the Jacobian of $f(x)$, which is a nonzero polynomial since the source Y has full rank. Let $\text{Sing} = \{x \in \mathbb{F}^k \mid J(x) = 0\}$ be the set of singular points and for each $a = a^{(-i)} \in \mathbb{F}^{k-1}$ let $\text{Sing}_a = \text{Sing} \cap V_a$.

Definition 4.8. *We say that $a = a^{(-i)} \in \mathbb{F}^{k-1}$ is "good" if it satisfies the following three conditions:*

1. $|V_a| \geq p^{5/6}$.
2. $|Sing_a| \leq p^{1/6}$.
3. $\hat{V}_a$ *is a curve. That is,* $\dim(\hat{V}_a) = 1$.

We define the set $S_i \subset \mathbb{F}^{k-1}$ to be the set of all "good" as.

The next claim shows that most as are "good". Thus proving part (1) of Lemma 4.10.

Claim 4.8.1. *Let S_i be as above. Then $\Pr[f^{(-i)} \in S_i] \geq 1 - p^{-\Omega(1)}$, where the probability is over uniformly chosen $x \in \mathbb{F}^k$.*

Proof. Let $a = a^{(-i)} \in \mathbb{F}^{k-1}$ be the random variable sampled by $a = f^{(-i)}(x)$, x uniform. For $1 \leq j \leq 3$ let E_j denote the event that a satisfies condition j in Definition 4.8. We can write

$$\Pr[a \text{ is "bad"}] \leq \Pr[E_1^c] + \Pr[E_2^c] + \Pr[E_1 \wedge E_2 \wedge E_3^c]. \tag{4.6}$$

We will bound each of these three probabilities independently by $p^{-\Omega(1)}$, which will prove the claim. The first probability can be seen to be bounded by $p^{-1/6}$ by a simple union bound on all as with small $|V_a|$.

To bound the second probability we first observe that $|\mathrm{Sing}| \leq \deg(J(x)) \cdot p^{k-1} \leq dk \cdot p^{k-1}$. Therefore, the number of different as not satisfying condition (2) is at most $dk \cdot p^{k-7/6}$. From Theorem 4.1 we have that for every $a = a^{(-i)} \in \mathbb{F}^{k-1}$ the set V_a contains at most $d^k \cdot p$ non-singular points. Therefore, the size of the union of all V_as for which condition (2) is not satisfied is bounded by

$$kd \cdot p^{k-1} + (kd \cdot p^{k-7/6})(d^k \cdot p) \leq p^{k-\Omega(1)}$$

(the first term counts all singular points and the second term counts all nonsingular points), where the inequality holds for $p > d^{Ck}$ for sufficiently large constant C. Therefore the second probability in Eq. 4.6 is also bounded by $p^{-\Omega(1)}$.

We now bound the third probability in Eq. 4.6. Let $A \subset \mathbb{F}^{k-1}$ be the set of as satisfying conditions (1) and (2) but not (3) in the definition of a "good" a. We first observe that Lemma 4.9 gives us the bound $|A| \leq kd^k \cdot p^{k-2}$ on the size of A. Now, for each $a \in A$ the size of V_a is bounded by $p^{1/6} + d^k \cdot p$ (V_a does not contain many singular points since a satisfies condition (2)). Therefore, we have that

$$\sum_{a \in A} |V_a| \leq |A| \cdot (p^{1/6} + d^k \cdot p) \leq kd^k \cdot p^{k-2} \cdot (p^{1/6} + d^k \cdot p) \leq p^{k-\Omega(1)}$$

(when $p > d^{Ck}$ and C is sufficiently large). This completes the proof of the claim. □

We now move to proving part (2) of Lemma 4.10. We will show that for every $a = a^{(-i)} \in S_i$ and for every nontrivial character χ the sum $|\Upsilon_\chi(a)|$ is bounded by $p^{-\Omega(1)}$.

Claim 4.8.2. *Let $a = a^{(-i)} \in S_i$. Then we have the bound $|\Upsilon_\chi(a)| \leq p^{-\Omega(1)}$.*

Proof. Let $\hat{V}_a = \hat{C}_1 \cup \ldots \cup \hat{C}_L$ be the decomposition of the curve $\hat{V}_a$ into irreducible components and let $C_j = \hat{C}_j \cap \mathbb{F}^k$ for $j \in [L]$. From Lemma B.12 we have that $L \leq d^k$. We wish to use Theorem 4.6 to bound $|\Upsilon_\chi(a)|$. Our first step will be to show that the polynomial $f_i(x)$ can be constant only on those irreducible components $\hat{C}_j$ that have few points in $\mathbb{F}_p$. To show this, notice that if the polynomial $f_i(x)$ is constant on one of the irreducible components $\hat{C}_j$ then, using Theorem 4.1 and part (2) of the definition of "good" *a*s, we get that $|C_j| \leq p^{1/6} + d^k$.

We now consider the modified curve $\hat{U}_a$ constructed by taking the union of those components $\hat{C}_j$ of $\hat{V}_a$ for which $|C_j| > p^{1/6} + d^k$ and let $U_a = \hat{U}_a \cap \mathbb{F}^k$. We can now use Theorem 4.6 to get the bound

$$\left| \sum_{x \in U_a} \chi(f_i(x)) \right| \leq 4d^{2k} \cdot p^{1/2},$$

which translates into the bound

$$\left| \sum_{x \in V_a} \chi(f_i(x)) \right| \leq d^k \cdot (p^{1/6} + d^k) + 4d^{2k} p^{1/2} \leq p^{2/3}$$

(separating the sum into points in the small components and in the large components) where the inequality hold when $p > d^{Ck}$, C sufficiently large. Dividing this sum by $|V_a| > p^{5/6}$ we get the required bound of $p^{-\Omega(1)}$ on $|\Upsilon_\chi(a)|$. □

Combining the above two claims concludes the proof of Lemma 4.10. □

4.6 Improving the Output Length

The extractor constructed in Section 4.5 can extract a constant fraction of the min-entropy of the source. It was suggested to us by Salil Vadhan that we can extract almost all of the min-entropy by using special properties of the source. This indeed works, and in this section we explain how.

We recall the notations of the last section: let $Y : \mathbb{F}^k \to \mathbb{F}^k$ be a (k, k, d)-polynomial source. Before describing the improved construction we need to define *seeded extractors*. For this section only we denote by U_s the uniform distribution on s bits.

Definition 4.9. *A function $E : \{0,1\}^n \times \{0,1\}^s \to \{0,1\}^m$ is an (r, ϵ)-*seeded extractor *if for every distribution X such that $H_\infty(X) \geq r$ the distribution $E(X, U_s)$ is ϵ-close to uniform. E is said to be* explicit *if it can be computed in polynomial time.*

Roughly speaking the method to extract many bits from Y is as follows: Let $E_1 : \mathbb{F} \to \{0,1\}^{m_1}$ be the extractor for distributions with small Fourier coefficients given by Lemma 4.8 (namely the $\mod 2^{m_1}$ function) and let $E_2 : \mathbb{F}^{k-1} \times \{0,1\}^s \to \{0,1\}^{m_2}$ be any seeded extractor with seed length s and output length m_2. Consider the composition of these two extractors given by $E(Y) = E_2(Y^{(-k)}, E_1(Y_k))$ (recall that $Y^{(-k)} = (Y_1, \ldots, Y_{k-1})$), in which the role of the uniform seed is taken by $E_1(Y_k)$. We would like to claim that $E(Y)$ is close to uniform. The first thing to observe is that m_1 has to be larger than s. This requirement will be easy to satisfy since in our setting, when $p \geq d^{O(k)}$, the output of E_1 will be larger than the seed length of standard seeded extractors. The more important thing to justify is the fact that we can replace the uniform seed of E_2 with a seed that is correlated with the source, $Y^{(-k)}$. This can be done since for 'most' fixings of $Y^{(-k)}$, the random variable $E_1(Y_k)$ is close to uniform (this follows from Bombieri's Theorem and the analysis of Section 4.5). We formalize this intuition in the following theorem:

Theorem 4.7. *Let k, d be integers and let $\mathbb{F}$ be a prime field of size $p > d^{\Omega(k)}$. Let $m_1 = c \cdot \log(p)$ for some small absolute constant c. Let $E_1 : \mathbb{F} \to \{0,1\}^{m_1}$ be the function computing $E_1(j) = \mod_{2^{m_1}}(j)$ and let $E_2 : \mathbb{F}^{k-1} \times \{0,1\}^s \to \{0,1\}^{m_2}$ be an (r, ϵ)-seeded extractor.*[1] *Suppose that $m_1 \geq s$ and $r \leq (k-1) \cdot \log\left(\frac{p}{2d}\right)$. Then, for any (k,k,d)-polynomial source $Y : \mathbb{F}^k \to \mathbb{F}^k$ we have that $E_2(Y^{(-k)}, E_1(Y_k))$ is ϵ'-close to uniform, with $\epsilon' = \epsilon + p^{-\Omega(1)}$ (we will use the convention that if $m_1 > s$ then E_2 uses only the first s bits of $E_1(Y_k)$).*

Proof. Assume w.l.o.g that $m_1 = s$. Using Lemma 4.10 together with Lemma 4.8 we get that with probability at least $1 - p^{-\Omega(1)}$ over a random fixing $Y^{(-k)} = b^{(-k)}$, the distribution $\left(E_1(Y_k) | Y^{(-k)} = b^{(-k)}\right)$ is $p^{-\Omega(1)}$-close to uniform. This means that the joint distribution $(Y^{(-k)}, E_1(Y_k))$ is $p^{-\Omega(1)}$-close to $(Y^{(-k)}, U_s)$. Therefore, we have that $E_2(Y^{(-k)}, E_1(Y_k))$ is $p^{-\Omega(1)}$-close to $E_2(Y^{(-k)}, U_s)$, which is $\epsilon + p^{-\Omega(1)}$ close to uniform by the properties of E_2. Here we use the fact that $r \leq (k-1) \cdot \log\left(\frac{p}{2d}\right)$ and that, from Lemma 4.3, $Y^{(-k)}$ is $p^{-\Omega(1)}$-close to having min-entropy at least $(k-1) \cdot \log\left(\frac{p}{2d}\right)$. □

Applying the last theorem with an appropriate seeded extractor enables us to construct a deterministic extractor for polynomial sources that extracts any constant fraction of the entropy of the source. It is possible to increase further the output length by using different seeded extractors. However, using current state-of-the-art seeded extractors, this would cost in terms of the error of the final construction. In order to avoid these complications we concentrate on extracting only a constant fraction (arbitrarily close to 1) of the min-entropy.

Theorem 4.8. *Let k and $d > 1$ be integers and let $\mathbb{F}$ be a field of prime cardinality $p > d^{\Omega(k)}$. Let $0 < \alpha < 1$. Then, there exists a function $E : \mathbb{F}^k \to \{0,1\}^m$ that is an explicit (k, d, ϵ)-extractor for polynomial sources over $\mathbb{F}^k$ with $m = (1-\alpha) \cdot k \cdot \log\left(\frac{p}{2d}\right)$ and $\epsilon = p^{-\Omega(1)}$.*

[1] We can safely ignore the technicality that p^{k-1} is not a power of 2.

Proof. We use the seeded extractors of [54] in conjunction with Theorem 4.7. In [54] it is shown that there exists an explicit (r, ϵ)-seeded extractor $E_2 : \mathbb{F}^{k-1} \times \{0,1\}^s \to \{0,1\}^{m_2}$ with the following parameters:

$$r = \lfloor (k-1) \cdot \log\left(\frac{p}{2d}\right) \rfloor,$$

$$\epsilon = p^{-\Omega(1)},$$

$$m_2 \geq (1 - \alpha/2) \cdot r,$$

$$s = O(\log^2(k \cdot \log(p)) + \log(1/\epsilon)) = O(\log(p)).$$

Plugging E_2 into the setting described in Theorem 4.7 we get an extractor with output length $m_2 \geq (1 - \alpha/2)(k-1) \cdot \log\left(\frac{p}{2d}\right)$, which is larger than $(1-\alpha) \cdot k \cdot \log\left(\frac{p}{2d}\right)$. □

4.7 Extractors for Weak Polynomial Sources

In this section we discuss the more general class of sources defined in the introduction as (n, k, d)-weak polynomial sources. Our final goal will be to prove Theorem 4.4, which we restate here for convenience:

Theorem 4.4. *There exist absolute constants C and c such that the following holds: Let $k \leq n$ and $d > 1$ be integers and let $d' = 8k^2d^3n$. Let $\mathbb{F}$ be a field of prime cardinality $p > (d')^{Ck}$. Then, there exists a function $E : \mathbb{F}^n \to \{0,1\}^m$ that is an explicit (k, d, ϵ)-extractor for weak polynomial sources over $\mathbb{F}^n$ with $m = \lfloor c \cdot k \cdot \log(p) \rfloor$ and $\epsilon = p^{-\Omega(1)}$.*

Theorem 4.4 will be a simple corollary of the following theorem, which shows that any (n, k, d)-WPS is close to a convex combination of (n, k, d)-polynomial sources.

Theorem 4.9. *Let $\mathbb{F}$ be a field of prime cardinality p. Let $k \leq n$ and d be integers such that $p > \max\{4D^2, 2^{10}\}$, where $D = (2k+1)d^{2k}$. Let X be an (n, k, d)-WPS over $\mathbb{F}$. Then X is δ-close to a convex combination of (n, k, d)-polynomial sources over $\mathbb{F}$, with $\delta = \frac{d \cdot k}{p}$.*

Before proving Theorem 4.9 we show how it can be used to prove Theorem 4.4.

Proof of Theorem 4.4. Let X be an (n, k, d)-WPS. We take the extractor $E : \mathbb{F}^k \to \{0,1\}^m$ to be the one given by Corollary 4.1 (namely, the extractor for polynomial sources). Using Theorem 4.9 we get that X is δ-close to a convex combination of (n, k, d)-polynomial sources, with $\delta = \frac{d \cdot k}{p} = p^{-\Omega(1)}$ (when $p > (d')^{Ck}$ and C is sufficiently large). We know from Corollary 4.1 that E is a (k, d, ϵ)-extractor for polynomial sources over $\mathbb{F}^n$, with $\epsilon = p^{-\Omega(1)}$. Therefore, $E(X)$ is δ-close to a convex combination of distributions, each of which is ϵ-close to uniform. It follows, using standard probability theory, that $E(X)$ is $(\delta + \epsilon) = p^{-\Omega(1)}$-close to uniform.

4.7.1 Proof of Theorem 4.9

The proof of the theorem will be in two steps. The first step will be to show that every (n,k,d)-WPS is sampled by a mapping $x : \mathbb{F}^n \to \mathbb{F}^n$ such that $\mathrm{rank}(x) \geq k$. The second step will be to show that a distribution sampled by such a mapping is close to a convex combination of (n,k,d)-polynomial sources. The first step of the proof of Theorem 4.9 is given by the following lemma.

Lemma 4.11. *Let $\mathbb{F}$ be a field of prime cardinality p. Let $k \leq n$ and d be integers such that $p \geq \max\{4D^2, 2^{10}\}$, where $D = (2k+1)\cdot d^{2k}$. Let X be an (n,k,d)-WPS over $\mathbb{F}$. Then there exists a mapping $x \in \mathcal{M}(\mathbb{F}^n \to \mathbb{F}^n)$ with rank $\geq k$ such that $X = x(U_n)$.*

The main thing that is needed in order to prove Lemma 4.11 is to show that if a polynomially sampled distribution has high entropy, then its rank is also high. In other words, we need to show that if the rank is low, so is the entropy. We achieve this kind of bound in two parts. The first part bounds the entropy of the output distribution of k *dependent* polynomials, that is, of k polynomials with rank at most $k-1$. This can be viewed as the "base case" for the proof of Lemma 4.11.

Lemma 4.12. *Let $\mathbb{F}$ be a field of prime cardinality p. Let k, n and d be integers such that $p > D$, where $D = (n+1)d^n$. Let $f_1, \ldots, f_k \in \mathbb{F}[x_1, \ldots, x_n]$ be k algebraically dependent polynomials of total degree at most d. Let P denote the distribution of the mapping $f = (f_1, \ldots, f_k) : \mathbb{F}^n \to \mathbb{F}^k$ on a uniformly chosen input in $\mathbb{F}^n$. Then P has support size at most $D \cdot p^{k-1}$.*

Proof. From Theorem 4.3 we know that there exists a nonzero polynomial $h \in \mathbb{F}[z_1, \ldots, z_k]$ of degree $\leq D$ such that $h(f_1(x), \ldots, f_k(x)) \equiv 0$ (notice that we use Theorem 4.3 with the roles of k and n reversed). Therefore, the support of P is contained in the zero set of h, whose size is bounded by $D \cdot p^{k-1}$ by Schwartz-Zippel (Lemma 4.3). □

The second auxiliary lemma we will need in the proof of Lemma 4.11 is the following lemma, which will enable us to reduce the number of variables of a mapping (assuming the number of variables is considerably larger than the number of outputs) while maintaining both the rank and the overall entropy of the mapping.

Lemma 4.13. *Let $\mathbb{F}$ be a finite field of cardinality q. Let d, k, n, m be integers such that $2k \leq n$. Let $x \in \mathcal{M}(\mathbb{F}^n \to \mathbb{F}^m, d)$ be such that $H_\infty(x(U_n)) \geq k \cdot \log(q)$. Then, there exists an affine subspace $V \subset \mathbb{F}^n$ of dimension $2k$ such that the restriction of x to V has min-entropy at least $k \cdot \log(q) - 2$. That is, if we denote by U_V the uniform distribution on V, then we have $H_\infty(x(U_V)) \geq k \cdot \log(q) - 2$.*

Proof. Take V to be a random affine subspace of dimension $2k$. For each $y \in \mathbb{F}^m$ let $S_y \triangleq \{t \in \mathbb{F}^n \mid x(t) = y\}$ and let $r_y \triangleq |S_y| \cdot q^{-n} = \Pr[x(U_n) = y]$. Fix some $y \in \mathbb{F}^m$. The expectation, over the choice of V, of $|S_y \cap V|$ is $q^{2k} \cdot r_y$. We can also bound the variance of $|S_y \cap V|$ (using pairwise independence of the points on V) by $|S_y|q^{2k-n}(1 - q^{2k-n}) \leq q^{2k} \cdot r_y$. Applying Chebyshev's inequality, and using the fact that for all $y \in F^m$ we have $r_y \leq q^{-k}$, one can show that

$$\Pr_V[\,|S_y \cap V| > 4q^k\,] \leq \frac{r_y}{9}. \tag{4.7}$$

Using the union bound we get that the probability that there exists a y for which the event in (4.7) happens is bounded by 1/9, and so there exists V such that for all $y \in F^m$ we have $|S_y \cap V| \leq 4q^k$. This completes the proof of the lemma since

$$\Pr[x(U_V) = y] = \frac{|S_y \cap V|}{q^{2k}} \leq 4q^{-k}.$$

□

The third auxiliary lemma we will use in the proof of Lemma 4.11 is the following one, which enables us to reduce the number of polynomials from n to k while maintaining most of the entropy.

Lemma 4.14. *Let $\mathbb{F}$ be a finite field of cardinality q. Let $k \leq n$ be integers and let $0 < s \leq k$ be a real number. Let X be a distribution over $\mathbb{F}^n$ such that $H_\infty(X) \geq s \cdot \log(q)$. Then there exists a linear mapping $l : \mathbb{F}^n \to \mathbb{F}^k$ such that for every $\alpha > 0$ the distribution $l(X)$ is ϵ-close to having min-entropy $\geq (s - \alpha) \cdot \log(q)$, where $\epsilon = \sqrt{2} \cdot q^{-\alpha/2}$.*

Proof. Let $\mathcal{L}$ denote the set of all linear mappings from $\mathbb{F}^n$ to $\mathbb{F}^k$ and let L be a random variable uniformly distributed over $\mathcal{L}$. Let us observe the average collision probability of $l(X)$ when we average over all $l \in \mathcal{L}$.

$$\begin{aligned}
\frac{1}{|\mathcal{L}|}\sum_{l\in\mathcal{L}} \mathrm{cp}(l(X)) &= \sum_{l\in\mathcal{L}} \Pr[L = l] \cdot \Pr_{x_1,x_2 \leftarrow X}[L(x_1) = L(x_2) \mid L = l] \\
&= \Pr_{x_1,x_2 \leftarrow X}[L(x_1) = L(x_2)] \\
&\leq \Pr_{x_1,x_2 \leftarrow X}[x_1 = x_2] + \Pr_{x_1,x_2 \leftarrow X}[L(x_1) = L(x_2) \mid x_1 \neq x_2] \\
&\leq q^{-s} + q^{-k} \leq 2q^{-s},
\end{aligned}$$

where in the last inequality we used the fact that the min-entropy of X is at least $\log(q^s)$ and so $\mathrm{cp}(X) \leq q^{-s}$. Therefore, there exists $l \in \mathcal{L}$ such that $\mathrm{cp}(l(X)) \leq 2q^{-s}$. Let $\alpha > 0$ and let us use Lemma 4.2 with $a = \frac{q^\alpha}{2}$ and $b = q^{s-\alpha}$. We therefore have $\mathrm{cp}(l(X)) \leq \frac{1}{ab}$ and so, by the lemma, $l(X)$ is $(1/\sqrt{a})$-close to having min-entropy at least $\log(b) = (s - \alpha) \cdot \log(q)$. □

One more simple auxiliary claim we will require is the following claim.

Claim 4.9.1. *Let* $0 < \epsilon < 1/4$*. Let* X *be some distribution on some finite set* Γ*. Suppose that* X *is* ϵ*-close to a distribution with support size at most* M*. Then* X *is* $(1/4)$*-far from any distribution with min-entropy at least* $\log(2M)$*.*

Proof. Assume towards a contradiction that there exists a distribution Y on Γ such that $\mathrm{H}_\infty(Y) \geq \log(2M)$ and $X \overset{\delta}{\sim} Y$ with $\delta \leq 1/4$. From the assumption on X we know that there exists a set $A \subset \Gamma$ with $|A| \leq M$ such that $\Pr[X \in A] \geq 1-\epsilon$. We therefore have that $\Pr[Y \in A] \geq 1-\epsilon-\delta > 1/2$. Therefore, since $\Pr[Y = a] \leq 2^{-\log \mathrm{H}_\infty(Y)} \leq \frac{1}{2M}$, we get that $\Pr[Y \in A] \leq |A| \cdot \frac{1}{2M} \leq 1/2$, a contradiction. □

We are now ready to prove Lemma 4.11.

Proof of Lemma 4.11 Let $x = x(t) \in \mathcal{M}(\mathbb{F}^n \to \mathbb{F}^n, d)$ be a mapping such that $X = x(U_n)$. We will show that $\mathrm{rank}(x) \geq k$. Assume towards a contradiction that $\mathrm{rank}(x) < k$. Using Lemma 4.13 we can replace x with a new polynomial mapping $\tilde{x} \in \mathcal{M}(\mathbb{F}^m \to \mathbb{F}^n, d)$, with $m = \min(n, 2k)$, and such that (a) $rank(\tilde{x}) \leq rank(x) < k$ and (b) $\mathrm{H}_\infty(\tilde{x}(U_m)) \geq (k-1/4)\log(q)$. Let $\tilde{X}$ denote the output distribution of $\tilde{x}$.

Next, we use Lemma 4.14 with parameters $\alpha = 1/4$ and $s = k - 1/4$. We get that there exists a linear mapping $l : \mathbb{F}^n \to \mathbb{F}^k$ such that $l(\tilde{X})$ is ϵ-close to having min-entropy at least $(k-1/2) \cdot \log(p)$, where

$$\epsilon = \sqrt{2} \cdot p^{1/8} < 1/4,$$

where the last inequality uses the fact that $p > 2^{10}$.

Notice that the distribution $l(\tilde{X})$ is the output distribution of k *dependent* polynomials. To see this write $D = \frac{\partial x}{\partial t}$ and let A_l be a $k \times n$ matrix representing l. The partial derivative matrix of $l \circ x$ is simply $A_l \cdot D$ and the rank of this matrix is at most the rank of D, which we assumed is bounded by $k-1$. Theorem 4.3 now implies that the polynomials sampling $l(\tilde{X})$ are dependent.

We can now use Lemma 4.12 to get that $l(\tilde{X})$ has support size at most $D \cdot p^{k-1}$, where $D = (m+1)d^m$. Therefore, by Claim 4.9.1, $l(\tilde{X})$ is $(1/4)$-far from any distribution with min-entropy at least $\log(2D \cdot p^{k-1})$. This implies

$$p^{k-1/2} < 2D \cdot p^{k-1},$$

which gives $p < 4D^2$, a contradiction. □

The second step in the proof of Theorem 4.9 is the following lemma.

Lemma 4.15. *Let* $\mathbb{F}$ *be a finite field. Let* $k \leq n$ *and* d *be integers. Let* $x \in \mathcal{M}(\mathbb{F}^n \to \mathbb{F}^n, d)$ *be a mapping with rank* k*. Let* X *be the distribution* $x(U_n)$*. Then* X *is* ϵ*-close to a convex combination of* (n,k,d)*-polynomial sources over* $\mathbb{F}$*, where* $\epsilon = \frac{d \cdot k}{|\mathbb{F}|}$*.*

Proof. Denote by D the submatrix of the first k rows and k columns of $\frac{\partial x}{\partial t}$, i.e.,

$$D = \begin{pmatrix} \frac{\partial x_1}{\partial t_1} & \cdots & \frac{\partial x_1}{\partial t_k} \\ \vdots & \ddots & \vdots \\ \frac{\partial x_k}{\partial t_1} & \cdots & \frac{\partial x_k}{\partial t_k} \end{pmatrix}.$$

We can assume w.l.o.g that D is non singular (this can be obtained by relabelling the ts and xs). Let $f : \mathbb{F}^n \to \mathbb{F}$ be defined as $f(t) \triangleq \det(D)(t)$. By assumption, f is nonzero and $deg(f) \leq d \cdot k$. For $c = (c_{k+1}, \ldots, c_n) \in \mathbb{F}^{n-k}$ define the mapping $x_c : \mathbb{F}^k \to \mathbb{F}^n$ as x restricted to c, that is, $x_c(t_1, \ldots, t_k) \triangleq x(t_1, \ldots, t_k, c_{r+1}, \ldots, c_n)$. Note that, the first k rows of $\frac{\partial x_c}{\partial t}$ are exactly D under the restriction $t_{k+1} = c_{k+1}, \ldots, t_n = c_n$. Thus $\frac{\partial x_c}{\partial t}$ has full rank whenever $f_c(t_1, \ldots, t_k) \triangleq f(t_1, \ldots, t_k, c_{k+1}, \ldots, c_n)$ is nonzero. Using Claim 4.7.2, $f_c \equiv 0$ with probability at most $\frac{d \cdot k}{|\mathbb{F}|}$ (for uniformly chosen c). Let X_c be the distribution $x_c(U_k)$. Then X is a convex combination of the X_cs. Moreover, using Lemma 4.1, X is $\frac{d \cdot k}{|\mathbb{F}|}$-close to a convex combination of the X_cs for which f_c is nonzero, and these X_cs are (n, k, d)-polynomial sources over $\mathbb{F}$. □

Proof of Theorem 4.9 We first apply Lemma 4.11 to get that X is sampled by a rank k mapping $x : \mathbb{F}^n \to \mathbb{F}^n$. Then we use Lemma 4.15 to show that $X = x(U_n)$ is δ-close to a convex combination of (n, k, d)-polynomial sources with $\delta = \frac{d \cdot k}{p}$. □

4.7.2 The Entropy of a Polynomial Mapping

We can use the results of the last section to show that, over sufficiently large fields, the entropy of a distribution sampled by a low-degree mapping $x \in \mathcal{M}(\mathbb{F}^n \to \mathbb{F}^n, d)$ is always "close" to $k \cdot \log(p)$, where k is equal to the rank of x. This is a generalization of the affine case, where the entropy is *exactly* $k \cdot \log(p)$. This is stated formally by the following theorem.

Theorem 4.10. *Let $k \leq n$ and d be integers. Let $D = (2k+1)d^{2k}$ and let $0 < \delta < 1$ be a real number. Let $\mathbb{F}$ be a field of prime cardinality p such that $p > \max\{(2d)^{\frac{k}{\delta}}, 2^{\frac{10}{\delta}}, (2D)^{\frac{2}{\delta}}\}$. Let $x \in \mathcal{M}(\mathbb{F}^n \to \mathbb{F}^n, d)$ be of rank k and let $X = x(U_n)$ be the distribution sampled by x. Then*

1. *X has min-entropy $\leq (k + \delta) \cdot \log(p)$.*
2. *X is ϵ-close to having min-entropy $\geq (k - \delta) \cdot \log(p)$, where $\epsilon = \frac{2 \cdot d \cdot k}{p}$.*

Proof. We start with a proof of part 2, which is easier. We apply Lemma 4.15 to get that X is $\frac{d \cdot k}{p}$-close to a convex combination of (n, k, d)-polynomial sources. From Theorem 4.3 we have that every distribution in this convex combination is $\frac{d \cdot k}{p}$-close to having min-entropy $\geq k \cdot \log\left(\frac{p}{2d}\right)$. It follows that

X is $\frac{2 \cdot d \cdot k}{p}$-close to having min-entropy at least

$$k \cdot \log\left(\frac{p}{2d}\right) \geq (k - \delta) \cdot \log(p),$$

where the inequality follows from the bound $p \geq (2d)^{\frac{k}{\delta}}$.

We proceed to prove part 1 of the theorem. We can assume w.l.o.g that $k < n$, for otherwise an entropy upper bound of $n \cdot \log(p)$ would be trivial. Suppose for contradiction that $\mathrm{H}_\infty(x) > (k + \delta) \cdot \log(p)$. Using Lemma 4.13 we can replace x with a new polynomial mapping $\tilde{x} \in \mathcal{M}(\mathbb{F}^m \to \mathbb{F}^n, d)$, with $m = \min(n, 2k)$, and such that (a) $rank(\tilde{x}) \leq rank(x) = k$ and (b) $\mathrm{H}_\infty(\tilde{x}(U_m)) \geq (k + \frac{3}{4}\delta)\log(p)$, where we need to use the following inequality

$$(k + \frac{3}{4}\delta)\log(p) \leq (k + \delta)\log(p) - 2,$$

which holds for $p > 2^{\frac{10}{\delta}}$.

Let $\tilde{X}$ denote the output distribution of $\tilde{x}$. We apply Lemma 4.14 with $\alpha = \delta/4$ to find a linear mapping $l : \mathbb{F}^n \to \mathbb{F}^{k+1}$ such that $l(\tilde{X})$ is ϵ'-close to having min-entropy at least $(k + \delta/2) \cdot \log(p)$ with $\epsilon' = \sqrt{2} \cdot p^{-\delta/8} < 1/4$ (here we use again the bound $p > 2^{\frac{10}{\delta}}$).

We proceed in a similar manner as in the proof of Lemma 4.11: We first use Lemma 4.12 to claim that $l(\tilde{X})$ has support size at most $D \cdot p^k$, where $D = (m+1)d^m$ (again, using the fact that $l \circ \tilde{x}$ has rank at most $\mathrm{rank}(\tilde{x}) \leq k$). From this fact and from Claim 4.9.1 we deduce that

$$(k + \delta/2) \cdot \log(p) \leq \log(2D \cdot p^k),$$

which is a contradiction since $p > (2D)^{\frac{2}{\delta}}$. □

4.8 Rank Extractors over the Complex Numbers

In this section we discuss the interpretation of rank extractors over the complex numbers. This interpretation will follow from the results appearing in [21] on algebraic independence and full-rank mappings over $\mathbb{C}$. The following theorem shows that over the complex numbers algebraic independence is equivalent to full rank.

Theorem 4.11 (Theorem 2.3 in [21]). *Let $x \in \mathcal{M}(\mathbb{C}^k \to \mathbb{C}^r, d)$ where $r \leq k$. The mapping x has full rank, that is, rank r, if and only if $x_1, \ldots, x_r$ are algebraically independent.*

The next theorem shows that for a mapping $x \in \mathcal{M}(\mathbb{C}^k \to \mathbb{C}^k, d)$, full rank is equivalent to having an image that is "essentially all" of $\mathbb{C}^k$, more precisely, all of $\mathbb{C}^k$ except for a set of measure zero. The theorem follows immediately from Theorem 2.4 in [21].

Theorem 4.12. *Fix any integers d, k and any $x \in \mathcal{M}(\mathbb{C}^k \to \mathbb{C}^k, d)$. The mapping x has full rank if and only if the image $x(\mathbb{C}^k)$ of x contains all of $\mathbb{C}^k$ except a set $Z \subseteq \mathbb{C}^k$ of measure zero in $\mathbb{C}^k$.*

Proof. Assume that x has full rank. In the proof of Theorem 2.4 in [21], it is shown that $x(\mathbb{C}^k)$ contains all of $\mathbb{C}^k$ except the set Z of zeros of some polynomial $H : \mathbb{C}^k \to \mathbb{C}$. Such a set Z has measure zero. Now assume $x(\mathbb{C}^k)$ contains all of $\mathbb{C}^k$ except for a set of measure zero in $\mathbb{C}^k$. Then $x(\mathbb{C}^k)$ is dense in $\mathbb{C}^k$ and it follows from Theorem 2.4 in [21] that $x_1, \ldots, x_n$ are algebraically independent, and therefore by Theorem 4.11, x has full rank. □

It follows that our constructions of rank extractors can be viewed as "dispersers" for low-degree sources over $\mathbb{C}$. That is, they are fixed mappings that map every k-dimensional low-degree source over $\mathbb{C}^n$ into almost all of $\mathbb{C}^k$.

Corollary 4.4. *Fix any integers d, k and n with $n \geq k$. Let $y : \mathbb{C}^n \to \mathbb{C}^k$ be the mapping from Theorem 4.1. Then, for any $x \in \mathcal{M}(\mathbb{C}^k \to \mathbb{C}^n, d)$ with full rank, $y(x(\mathbb{C}^k))$ contains all of $\mathbb{C}^k$ except for a set of measure zero.*

As far as we know, this kind of generalized dispersers was not considered before, and it will be interesting to find applications for it.

4.9 Discussion and Open Problems

Our chapter invites further work in several directions.[2]

- The extractors we give in this chapter work when the field size is $d^{\Omega(k)}$. Extending our results to the case where the field size is polynomial in k is an interesting open problem. Building on the results of this chapter it is enough to construct such an extractor for polynomial sources of full rank.

- An affine source may be viewed in two dual ways: as the image of an affine map, or as the kernel of one. Our extension here to low-degree sources takes the first view. An interesting problem is extending the second view: extracting from low-degree algebraic varieties. We note that the case of one dimensional varieties is already covered by Bombieri's theorem (see Section 4.5).

- We prove an exponential upper bound of $(n+1)d^n$ on the degree of the annihilating polynomial for a set of degree d dependent polynomials in n variables. Can this bound be improved in general? Are there lower bounds? This seems to be open even over the complex numbers. An improvement on the upper bound above will yield a tighter connection between min-entropy and algebraic rank for smaller field sizes. However, it is possible that the latter can be obtained without the former.

---

[2]A recent work of Kayal [38] makes progress on several of these issues.

- What is the computational complexity of testing algebraic independence? When the field size affords the equivalence to the rank of the Jacobian, there is a simple RP algorithm. Can one do it for smaller fields?
- What is the complexity of finding an annihilating polynomial when the polynomials are dependent? Our degree bound guarantees a PSPACE algorithm. Is there a better one, or can this problem be complete for this class?

Chapter 5

Increasing the Output Length of Zero-Error Dispersers

Summary

Let $\mathcal{C}$ be a class of probability distributions over a finite set Ω. A function $D : \Omega \mapsto \{0,1\}^m$ is a *disperser* for $\mathcal{C}$ with *entropy threshold* k and *error* ϵ if for any distribution X in $\mathcal{C}$ such that X gives positive probability to at least 2^k elements we have that the distribution $D(X)$ gives positive probability to at least $(1-\epsilon)2^m$ elements. A long line of research is devoted to giving explicit (that is, polynomial-time computable) dispersers (and related objects called "extractors") for various classes of distributions while trying to maximize m as a function of k.

In this chapter we are interested in explicitly constructing *zero-error dispersers* (that is, dispersers with error $\epsilon = 0$). For several interesting classes of distributions there are explicit constructions in the literature of zero-error dispersers with "small" output length m, and we give improved constructions that achieve "large" output length, namely $m = \Omega(k)$.

We achieve this by developing a general technique to improve the output length of zero-error dispersers (namely, to transform a disperser with short output length into one with large output length). This strategy works for several classes of sources and is inspired by the transformation that improves the output length of extractors used in previous chapters. However, we stress that this technique is different, and in particular gives nontrivial results in the errorless case.

Using our approach we construct improved zero-error dispersers for the class of 2-*sources*. More precisely, we show that for any constant $\delta > 0$ there is a constant $\eta > 0$ such that

A. Gabizon, *Deterministic Extraction from Weak Random Sources*, Monographs in Theoretical Computer Science. An EATCS Series, DOI 10.1007/978-3-642-14903-0_5, © Springer-Verlag Berlin Heidelberg 2011

for sufficiently large n there is a poly-time computable function $D : \{0,1\}^n \times \{0,1\}^n \mapsto \{0,1\}^{\eta n}$ such that for any two independent distributions X_1, X_2 over $\{0,1\}^n$ such that both of them support at least $2^{\delta n}$ elements we get that the output distribution $D(X_1, X_2)$ has full support. This improves the output length of previous constructions by [4] and has applications in Ramsey Theory and in constructing certain data structures [24].

We also use our techniques to give explicit constructions of zero-error dispersers for bit-fixing sources and affine sources over polynomially large fields. These constructions improve the best known explicit constructions due to [52, 25] and achieve $m = \Omega(k)$ for bit-fixing sources and $m = k - o(k)$ for affine sources.
This chapter is based on [27]

5.1 Introduction

5.1.1 Randomness Extractors and Dispersers

We start with formal definitions of extractors and dispersers, as they are used in this chapter

Definition 5.1 (min-entropy and statistical distance). *Let Ω be a finite set. Recall that the* min-entropy *of a distribution X on Ω is defined by $H_\infty(X) = min_{x\in\Omega} \log_2 \frac{1}{\Pr[X=x]}$. For a class $\mathcal{C}$ of distributions on Ω we use $\mathcal{C}_k$ to denote the class of all distributions $X \in \mathcal{C}$ such that $H_\infty(X) \geq k$. We say that two distributions X, Y on Ω are* ϵ-close *if $\frac{1}{2}\sum_{w\in\Omega} |\Pr[X = w] - \Pr[Y = w]| \leq \epsilon$.*

When given a class $\mathcal{C}$ of distributions (which we call "sources") the goal is to design one function that refines the randomness of any distribution X in $\mathcal{C}$. An *extractor* produces a distribution that is (close to) uniform whereas a *disperser* produces a distribution with (almost) full support. A precise definition follows.

Definition 5.2 (extractors and dispersers). *Let $\mathcal{C}$ be a class of distributions on a finite set Ω.*

- *A function $E : \Omega \mapsto \{0,1\}^m$ is an* extractor *for $\mathcal{C}$ with* entropy threshold k *and* error $\epsilon > 0$ *if for every $X \in \mathcal{C}_k$, $E(X)$ is ϵ-close to the uniform distribution on $\{0,1\}^m$.*
- *A function $D : \Omega \mapsto \{0,1\}^m$ is a* disperser *for $\mathcal{C}$ with* entropy threshold k *and* error $\epsilon > 0$ *if for every $X \in \mathcal{C}_k$, $|\mathrm{Supp}(D(X))| \geq (1-\epsilon)2^m$ (where* $\mathrm{Supp}(Z)$ *denotes the support of the random variable Z).*

We remark that every extractor is in particular a disperser and that the notion of dispersers only depends on the support of the distributions in $\mathcal{C}$. A long line of research is concerned with designing extractors and dispersers for various classes of sources. For a given class $\mathcal{C}$ we are interested in designing extractors and dispersers with as small as possible entropy threshold k, as large as possible output length m and as small as possible error ϵ. (We remark that it easily follows that $m \le k$ whenever $\epsilon < 1/2$.)

It is often the case that the probabilistic method gives that a randomly chosen function E is an excellent extractor. (This is in particular true whenever the class $\mathcal{C}$ contains "not too many" sources.) However, most applications of extractors and dispersers require *explicit constructions*, namely functions that can be computed in time polynomial in their input length. Much of the work done in this area can be described as an attempt to match the parameters obtained by existential results using explicit constructions.

5.1.2 Zero-Error Dispersers

In this work we are interested in *zero-error dispersers*. These are dispersers where the output distribution has full support. That is, for every source X in the class $\mathcal{C}$,

$$\{D(x) : x \in \mathrm{Supp}(X)\} = \{0,1\}^m.$$

We also consider a stronger variant which we call *strongly-hitting disperser*, in which every output element $z \in \{0,1\}^m$ is obtained with "not too small" probability. A precise definition follows.

Definition 5.3 (Zero-error dispersers and strongly hitting dispersers). *Let $\mathcal{C}$ be a class of distributions on a finite set Ω.*

- *A function D is a* zero-error disperser *for $\mathcal{C}$ with* entropy threshold k *if it is a disperser for $\mathcal{C}$ with entropy threshold k and error $\epsilon = 0$.*
- *A function $D : \Omega \mapsto \{0,1\}^m$ is a μ-*strongly hitting disperser *for $\mathcal{C}$ with* entropy threshold k *if for every $X \in \mathcal{C}_k$ and for every $z \in \{0,1\}^m$,* $\Pr[D(X) = z] \ge \mu$.

Note that a μ-strongly hitting disperser with $\mu > 0$ is in particular a zero-error disperser and that any μ-strongly hitting disperser has $\mu \le 2^{-m}$. The following facts immediately follow:

Fact 5.1. *Let $f : \Omega \mapsto \{0,1\}^m$ be a function and let $\epsilon \le 2^{-(m+1)}$.*

- *If f is a disperser with error ϵ then f is a zero-error disperser (for the same class $\mathcal{C}$ and entropy threshold k).*
- *If f is an extractor with error ϵ then f is a $2^{-(m+1)}$-strongly hitting disperser (for the same class $\mathcal{C}$ and entropy threshold k).*

It follows that extractors and dispersers with small ϵ immediately translate into zero-error dispersers (as one can truncate the output length to $m' = \log(1/\epsilon) - 1$ bits, and such a truncation preserves the output guarantees of extractors and dispersers).

5.1.3 Increasing the Output Length of Zero-Error Dispersers

For several interesting classes of sources there are explicit constructions of dispersers with "large" error (which by Fact 5.1 give zero-error dispersers with "short" output length). In this chapter we develop techniques to construct zero-error dispersers with large output length.

The composition approach

The following methodology for increasing the output length of extractors was suggested in [26, 63]: When given an extractor E' with "small" output length t (for some class $\mathcal{C}$), consider the function $E(x) = F(x, E'(x))$ where F is a seeded extractor. Shaltiel [63] (building on earlier work by Gabizon et al. [26]) shows that if E' and F fulfill certain requirements then this construction yields an extractor for $\mathcal{C}$ with large output length. The high-level idea is that if certain conditions are fulfilled, then the distribution $F(X, E(X))$ (in which the two inputs of F are *dependent*) is close to the distribution $F(X, Y)$ (where Y is an independent uniformly distributed variable); note that the latter distribution is close to uniform by the definition of seeded extractors. This technique proved useful for several interesting classes of sources.

We would like to apply an analogous idea to obtain zero-error dispersers. However, by the lower bounds of [48, 49], if F is a seeded extractor (or seeded disperser) then its seed length is at least $\log(1/\epsilon)$. This means that if we want $F(X, Y)$ to output m bits with error $\epsilon < 1/2^m$, we need seed length larger than m. This in turn means that we want E' to have output length $t > m$, which makes the transformation useless.

There are also additional problems. The argument in [63] requires the "original function" E' to be an extractor (and it does not go through if E' is a disperser), and furthermore the error of the "target function" E is at least as large as that of the "original function" E' (and once again we don't gain when shooting for zero-error dispersers).

Summing up we note that if we want to improve the output length of a zero-error disperser D' by a composition of the form $D(x) = F(x, D'(x))$, we need to use a function F with different properties (a seeded extractor or disperser will not do) and we need to use a different kind of analysis.

Composing zero-error dispersers

In this chapter we imitate the method of [63] and give a general method to increase the output length of zero-error dispersers. That is, when given:

- a zero-error disperser $D' : \Omega \mapsto \{0,1\}^t$ for a class $\mathcal{C}$ and "small" output length t and
- a function $F : \Omega \times \{0,1\}^t \mapsto \{0,1\}^m$ for "large" output length m,

we identify properties of F that are sufficient so that the construction

$$D(x) = F(x, D'(x))$$

gives a zero-error disperser. (The argument is more general and transforms $2^{-(t+O(1))}$-strongly hitting dispersers into $2^{-(m+O(1))}$-strongly hitting dispersers.) We then use this technique to give new constructions of zero-error dispersers and strongly-hitting dispersers.

Subsource hitters

As explained earlier we cannot choose F to be a seeded extractor. Instead, we introduce a new object which we call a *subsource hitter*. The definition of subsource hitters is somewhat technical and is tailored so that the construction $D(x) = F(x, D'(x))$ indeed produces a disperser.

Definition 5.4 (subsource hitter). *A distribution X' on Ω is a* subsource *of a distribution X on Ω if there exist $\alpha > 0$ and a distribution X'' on Ω such that X can be expressed as a convex combination $X = \alpha X' + (1 - \alpha)X''$.*

Let $\mathcal{C}$ be a class of distributions on Ω. A function $F : \Omega \times \{0,1\}^t \mapsto \{0,1\}^m$ is a subsource hitter *for $\mathcal{C}$ with* entropy threshold *k and* subsource entropy *$k - v$ if for any $X \in \mathcal{C}_k$ and $z \in \{0,1\}^m$ there exist a $y \in \{0,1\}^t$ and a distribution $X' \in \mathcal{C}_{k-v}$ that is a subsource of X such that for every $x \in \text{Supp}(X')$ we have that $F(x, y) = z$.*

A subsource hitter has the property that for any $z \in \{0,1\}^m$ there exist $y \in \{0,1\}^t$ and $x \in \text{Supp}(X)$ such that $F(x, y) = z$ and in particular

$$\{F(x, y) : x \in \text{Supp}(X), y \in \{0,1\}^t\} = \{0,1\}^m.$$

In addition, a subsource hitter has the stronger property that there exists a subsource X' of X (which is itself a source in $\mathcal{C}$) such that for any $z \in \{0,1\}^m$ there exists $y \in \{0,1\}^t$ such that for *any* $x \in \text{Supp}(X') \subseteq \text{Supp}(X)$, $F(x, y) = z$.

This property allows us to show that $D(x) = F(x, D'(x))$ is a zero-error disperser with entropy threshold k whenever D' is a zero-error disperser with entropy threshold $k - v$. This is because when given a source $X \in \mathcal{C}_k$ and $z \in \{0,1\}^m$ we can consider the seed $y \in \{0,1\}^t$ and subsource X' guaranteed in the definition. We have that D' is a zero-error disperser and that X' meets the entropy threshold of D'. It follows that there exists $x \in \text{Supp}(X') \subseteq \text{Supp}(X)$ such that $D'(x) = y$. It follows that

$$D(x) = F(x, D'(x)) = F(x, y) = z,$$

and this means that D indeed outputs z. (We remark that a more complicated version of this argument shows that the composition applies to strongly hitting dispersers). The exact details are given in Section 5.3.

It is interesting to note that this argument is significantly simpler than that of [63]. Indeed, the definition of subsource hitters is specifically tailored to make the composition argument go through and the more complicated task is to design subsource hitters. This is in contrast to [63] in which the function F is in most cases an "off-the-shelf" seeded extractor and the difficulty is to show that the composition succeeds.

5.1.4 Applications

We use the new composition technique to construct zero-error dispersers with large output length for various classes of sources. We discuss these constructions and some applications below.

Zero-error 2-source dispersers

The class of 2-*sources* is the class of distributions $X = (X_1, X_2)$ on $\Omega = \{0,1\}^n \times \{0,1\}^n$ such that X_1, X_2 are independent. It is common to consider the case where each of the two distributions X_1, X_2 has min-entropy at least some threshold k.

Definition 5.5 (2-source extractors and dispersers). *A function $f : \{0,1\}^n \times \{0,1\}^n \mapsto \{0,1\}^m$ is a 2-source extractor (or disperser) with entropy threshold $2 \cdot k$ and error $\epsilon \geq 0$ if for every two independent distributions X_1, X_2 on $\{0,1\}^n$ both having min-entropy at least k, $f(X_1, X_2)$ is ϵ-close to the uniform distribution on $\{0,1\}^m$ (or $|\text{Supp}(f(X_1, X_2))| \geq (1-\epsilon)2^m$). We say that f is a zero-error disperser if it is a disperser with error $\epsilon = 0$. We say that f is a μ-strongly hitting disperser if for every X_1, X_2 as above and every $z \in \{0,1\}^m$, $\Pr[f(X_1, X_2) = z] \geq \mu$.*

Background. The probabilistic method gives 2-source extractors with $m = 2 \cdot k - O(\log(1/\epsilon))$ for any $k \geq \Omega(\log n)$. However, until 2005 the best explicit constructions [14, 70] only achieved $k > n/2$. The current best extractor construction [9] achieves entropy threshold $k = (1/2 - \alpha)n$ for some constant $\alpha > 0$. Improved constructions of dispersers for entropy threshold $k = \delta n$ (for an arbitrary constant $\delta > 0$) were given in [4]. These dispersers can output any constant number of bits with zero error (and are μ-strongly hitting for some constant $\mu > 0$).[1] Subsequent work by [5] achieved entropy threshold to $k = n^{o(1)}$ and gives zero-error dispersers that output one bit.

Our results. We use our composition techniques to improve the output length in the construction of [4]. We show that:

[1] In [50] it is pointed out that by enhancing the technique of [4] using ideas from [5] and replacing some of the components used in the construction with improved components that are constructed in [50] it is possible to increase the output length and achieve a zero-error disperser with output length $m = k^{\Omega(1)}$ for the same entropy threshold k.

Theorem 5.6 (2-source zero-error disperser). *For every $\delta > 0$ there exists a $\nu > 0$ and an $\eta > 0$ such that for sufficiently large n there is a poly(n)-time computable $(\nu 2^{-m})$-strongly hitting 2-source disperser $D : (\{0,1\}^n)^2 \mapsto \{0,1\}^m$ with entropy threshold $2 \cdot \delta n$ and $m = \eta n$.*

Note that our construction achieves an output length that is optimal up to constant factors for this entropy threshold. For lower entropy threshold our technique gives that any explicit construction of a zero-error 2-source disperser D' with entropy threshold k and output length $t = \Omega(\log n)$ can be transformed into an explicit construction of a zero-error 2-source disperser D with entropy threshold $2 \cdot k$ and output length $m = \Omega(k)$. (See Section 5.4 for a precise formulation that also considers strongly hitting dispersers.) This cannot be applied on the construction of [5] that achieves entropy threshold $k = n^{o(1)}$ as this construction only outputs one bit. Nevertheless, this means that it suffices to extend the construction of [5] so that it outputs $\Theta(\log n)$ bits in order to obtain an output length of $m = \Omega(k)$ for low entropy threshold k.

We prove Theorem 5.6 by designing a subsource hitter for 2-sources and using our composition technique. The details are given in Section 5.4 and a high-level outline appears next.

Outline of the argument. We want to design a function $F : \{0,1\}^n \times \{0,1\}^n \times \{0,1\}^t \mapsto \{0,1\}^m$ such that for any 2-source $X = (X_1, X_2)$ with sufficient min-entropy and for any $z \in \{0,1\}^m$ there exist a "seed" $y \in \{0,1\}^t$ and a subsource X' of X such that $X' = (X'_1, X'_2)$ is a 2-source with roughly the same min-entropy as X and $\Pr[F(X'_1, X'_2, y) = z] = 1$. We will be shooting for $m = \Omega(n)$ and $t = O(\log n)$.

We construct the seed obtainer F using ideas from [4, 5]. Let E be a seeded extractor with seed length $t = O(\log n)$, output length $v = \Omega(k)$ and error $\epsilon_E = 1/100$ (such extractors were constructed in [42, 34]). When given inputs x_1, x_2, y we consider $r_1 = E(x_1, y)$ and $r_2 = E(x_2, y)$. By using a stronger variant of seeded extractors called "strong extractors" it follows that there exists a "good seed" $y \in \{0,1\}^t$ such that $R_1 = E(X_1, y)$ and $R_2 = E(X_2, y)$ are ϵ_E-close to uniform. We then use a 2-source extractor $H : \{0,1\}^v \times \{0,1\}^v \mapsto \{0,1\}^m$ for *very high* entropy threshold (say entropy threshold $2 \cdot 0.9v$) and very low error (say error $2^{-(m+1)}$ for output length $m = \Omega(v) = \Omega(k)$). Such extractors were constructed in [70]. Our final output is given by

$$F(x_1, x_2, y) = H(E(x_1, y), E(x_2, y)).$$

This seems strange at first sight as it is not clear why running H on inputs R_1, R_2 that are already close to uniform helps. Furthermore, the straightforward analysis only gives that $H(R_1, R_2)$ is ϵ-close to uniform for *large* error $\epsilon \geq \epsilon_E = 1/100$ and this means that the output of F may miss a large fraction of strings in $\{0,1\}^m$.

The point to notice is that both R_1, R_2 are close to uniform and therefore have large support $(1-\epsilon_E)2^v \geq 2^{0.9v}$. Using Fact 5.1 we can think of H as a zero-error disperser. Recall that dispersers are oblivious to the precise probability distribution of R_1, R_2 and it is sufficient that R_1, R_2 have large support. It follows that indeed every string $z \in \{0,1\}^m$ is hit by $H(R_1, R_2)$.

This does not suffice for our purposes as we need that any string z is hit with probability 1 on a subsource $X' = (X_1', X_2')$ of X in which the two distributions X_1' and X_2' are independent. For any output string $z \in \{0,1\}^m$ we consider a pair of values (r_1, r_2) for R_1, R_2 on which $H(r_1, r_2) = z$ (we have just seen that such a pair exists) and set $X_1' = (X_1 | E(X_1, y) = r_1)$ and $X_2' = (X_2 | E(X_2, y) = r_2)$. Note that these two distributions are indeed independent (as each depends only on one of the original distributions X_1, X_2) and that on every $x_1' \in \mathrm{Supp}(X_1')$ and $x_2' \in \mathrm{Supp}(X_2')$ we have that

$$F(x_1', x_2', y) = H(E(x_1', y), E(x_2', y)) = H(r_1, r_2) = z.$$

Furthermore, for a typical choice of (r_1, r_2) we can show that both X_1', X_2' have min-entropy roughly $k - v$. Thus, setting v appropriately, X' is a subsource of X with the required properties. (A more careful version of this argument can be used to preserve the "strongly hitting" property.)

Interpretation in Ramsey theory

A famous theorem in Ramsey Theory (see [32]) states that for sufficiently large N and any 2-coloring of the edges of the complete graph on N vertices there is an induced subgraph on $K = \Theta(\log N)$ vertices which is "monochromatic" (that is, all edges are of the same color).

Zero-error 2-source dispersers (with output length $m = 1$) can be seen as providing counterexamples to this statement for larger values of K in the following way: When given a zero-error 2-source disperser $D : \{0,1\}^n \times \{0,1\}^n \mapsto \{0,1\}^m$ with entropy threshold $2 \cdot k$ we can consider coloring the edges of the full graph on $N = 2^n$ vertices with 2^m colors by coloring an edge (v_1, v_2) by $D(v_1, v_2)$. (A technicality is that $D(v_1, v_2)$ may be different from $D(v_2, v_1)$, and to avoid this problem the coloring is defined by ordering the vertices according to some order and coloring the edge (v_1, v_2) where $v_1 \leq v_2$ by $D(v_1, v_2)$). The disperser guarantee can be used to show that any induced subgraph with $K = 2^{k+1}$ vertices contains edges of *all* 2^m colors.[2]

Note that dispersers with $m > 1$ translate into colorings with more colors and that in this context of Ramsey Theory the notion of a zero-error disperser seems more natural than one that allows error. Our constructions achieve $m = \Omega(k)$ and thus the number of colors in the coloring approaches the size of the induced subgraph.

---

[2] In fact, dispersers translate into a significantly stronger guarantee that discusses colorings of the edges of the complete N by N bipartite graph such that any induced K by K subgraph has all colors.

Generalizing this relation between dispersers and Ramsey theory we can view any zero-error disperser for a class $\mathcal{C}$ as a coloring of all $x \in \Omega$ such that any set S that is obtained as the support of a distribution in $\mathcal{C}$ is colored by all possible 2^m colors.

Rainbows and implicit $O(1)$-probe search

As we now explain, explicit constructions of zero-error 2-source dispersers can be used to construct certain data structures (this connection is due to [24]).

Consider the following problem. We are given a set $S \subseteq \{0,1\}^n$ of size 2^k. We want to store the elements of S in a table T of the same size, where every entry in the table contains a single element of S (and so the only freedom is in ordering the elements of S in the table T). We say that T supports q-queries if given $x \in \{0,1\}^n$ we can determine whether $x \in S$ using q queries to T (note for example that ordered tables and binary search support $q = k$ queries). Yao [75] and Fiat and Naor [24] showed that it is impossible to achieve $q = O(1)$ when n is large enough relative to k. (This result can be seen as a kind of Ramsey Theorem.)

Fiat and Naor [24] gave explicit constructions of tables that support $q = O(1)$ queries when $k = \delta \cdot n$ for any constant $\delta > 0$. This was achieved by reducing the implicit probe search problem to the task of explicitly constructing a certain combinatorial object that they call a "rainbow".

Loosely speaking, a rainbow is a zero-error disperser for the class of distributions X that are composed of q independent copies of a high min-entropy distribution. We stress that for this application one needs (strongly hitting) dispersers with large output length. More precisely, in order to support $q = O(1)$ queries one requires such dispersers that have output length m that is a *constant fraction* of the entropy threshold.

Our techniques can be used to explicitly construct rainbows which in turn allow implicit probe schemes that support $q = O(1)$ queries for smaller values of k than previously known. More precisely, for any constant $\delta > 0$ and $k = n^\delta$ there is a constant q and a scheme that supports q queries. The precise details are given in Section 5.4.5. (We remark that one can also achieve the same results by using the technique of [24] and plugging in recent constructions of seeded dispersers.)

Zero-error dispersers for bit-fixing sources

The class of *bit-fixing sources* is the class of distributions X on $\Omega = \{0,1\}^n$ such that there exists a set $S \subseteq [n]$ such that X_S (that is, X restricted to the indices in S) is uniformly distributed and $X_{[n]\setminus S}$ is constant. Note that for such a source X, $\mathrm{H}_\infty(X) = |S|$. (We remark that these sources are sometimes called "oblivious bit-fixing sources" to differentiate them from "non-oblivious bit-fixing sources" in which $X_{[n]\setminus S}$ is allowed to be a function of X_S.)

Background. The function $Parity(x)$ (that is, the exclusive-or of the bits of x) is obviously an extractor for bit-fixing sources with entropy threshold $k = 1$, error $\epsilon = 0$ and output length $m = 1$. It turns out that there are no errorless extractors for $m = 2$. More precisely, [15] showed that for $k < n/3$ there are no extractors for bit-fixing sources with $\epsilon = 0$ and $m = 2$. For larger values of k, [15] gives constructions with $m > 1$ and $\epsilon = 0$. For general entropy threshold k, the current best explicit construction of extractors for bit-fixing sources is due to [52] (in fact, this extractor works for a more general class of "low-weight affine sources"). These extractors work for any entropy threshold $k \geq (\log n)^c$ for some constant c, and achieve output length $m = (1 - o(1))k$ for error $\epsilon = 2^{-k^{\Omega(1)}}$. Using Fact 5.1 this gives a zero-error disperser with output length $m = k^{\Omega(1)}$.

Our results. We use our composition techniques to construct zero-error dispersers for bit-fixing sources with output length $m = \Omega(k)$. We show that:

Theorem 5.7 (Zero-error disperser for bit-fixing sources). *There exist $c > 1$ and $\eta > 0$ such that for sufficiently large n and $k \geq (\log n)^c$ there is a poly(n)-time computable zero-error disperser $D : \{0,1\}^n \mapsto \{0,1\}^m$ for bit-fixing sources with entropy threshold k and output length $m = \eta k$.*

Note that our construction achieves an output length that is optimal up to constant factors. We prove Theorem 5.7 by designing a subsource hitter for bit-fixing sources and using our composition technique. The details are given in Section 5.5 and a high-level outline appears next.

Outline of the argument. Our goal is to design a subsource hitter $G : \{0,1\}^n \times \{0,1\}^t \mapsto \{0,1\}^m$ for bit-fixing sources with entropy threshold k, output length $m = \Omega(k)$ and "seed length" $t = O(\log n)$. We make use of the subsource hitter for 2-sources $F : \{0,1\}^n \times \{0,1\}^n \times \{0,1\}^{O(\log n)} \mapsto \{0,1\}^m$ that we designed earlier. We apply it for entropy threshold $k' = k/8$ and recall that it has output length $m = \Omega(k') = \Omega(k)$.

When given a seed $y \in \{0,1\}^t$ for G we think about it as a pair of strings (y', y'') where y' is a seed for F and y'' is a seed for an explicit construction of pairwise independent variables $Z_1, \ldots, Z_n$ where for each i, Z_i takes values in $\{1,2,3\}$ (indeed there are such constructions with seed length $O(\log n)$). When given such a seed y'' we can use the values $Z_1, \ldots, Z_n$ to partition the set $[n]$ into three disjoint sets T_1, T_2, T_3 by having each index $i \in [n]$ belong to T_{Z_i}. We construct G as follows:

$$G(x, (y', y'')) = F(x_{T_1}, x_{T_2}, y').$$

In words, we use y'' to partition the given n-bit string into three strings and we run F on the first two strings (padding each of them to length n) using the seed y'.

We need to show that for any bit-fixing source X of min-entropy k and for any $z \in \{0,1\}^m$ there exist a seed $y = (y', y'')$ and a subsource X' of X such that X' is a bit-fixing source with roughly the same min-entropy as X and $\Pr[G(X', (y', y'')) = z] = 1$.

We have that X is a bit-fixing source and we let $S \subseteq [n]$ be the set of its "good indices". Note that $|S| \geq k$. By the "sampling properties" of pairwise independent distributions (see [31] for a survey on "averaging samplers") it follows that there exists a y'' such that for every $i \in [3]$, $|S \cap T_i| \geq k/8$. It follows that $X_{T_1}, X_{T_2}, X_{T_3}$ are bit-fixing sources with min-entropy at least $k/8$ (and note that these three distributions are independent). Thus, by the properties of the subsource hitter F there exist x_1, x_2, y' such that $F(x_1, x_2, y') = z$ (note that here we're only using the property that F "hits z" and do not use the stronger property that F "hits z on a subsource"). Consider the distribution

$$X' = (X | X_{T_1} = x_1 \wedge X_{T_2} = x_2).$$

This is a subsource of X which is a bit-fixing source with min-entropy at least $k/8$ (as we have not fixed the $k/8$ good bits in T_3). It follows that for every $x \in \text{Supp}(X')$

$$G(x, (y', y'')) = F(x_1, x_2, y') = z$$

and G is indeed a subsource hitter for bit-fixing sources.

Affine sources

The class of *affine sources* is the class of distributions X on $\Omega = \mathbb{F}_q^n$ (where $\mathbb{F}_q$ is the finite field of q elements) such that X is uniformly distributed over an affine subspace V in $\mathbb{F}_q^n$. Note that such a source X has min-entropy $\log q \cdot dim(V)$. Furthermore, any bit-fixing source is an affine source over $\mathbb{F}_2$.

Background. For $\mathbb{F}_2$ the best explicit construction of extractors for affine sources was given in [10]. This construction works for entropy threshold $k = \delta n$ (for any fixed $\delta > 0$) and achieves output length $m = \Omega(k)$ with error $\epsilon < 2^{-m}$.

Extractors for lower entropy thresholds were given by [25] in the case where $q = n^{\Theta(1)}$. For any entropy threshold $k > \log q$, these extractors can output $m = (1 - o(1))k$ bits with error $\epsilon = n^{-\Theta(1)}$. Using Fact 5.1 this gives zero-error dispersers with output length $m = \Theta(\log n)$.

Our results. Our composition techniques can be applied on affine sources. We focus on the case of large fields (as in that case we can improve the results of [25]). We remark that our techniques also apply when q is small (however, at the moment we do not gain by applying them on the existing explicit constructions). We prove the following theorem:

Theorem 5.8. *Fix any prime power q and integers n, k such that $q \geq n^{18}$ and $2 \leq k < n$. There is a $poly(n, \log q)$-time computable zero-error disperser $D : \mathbb{F}_q^n \mapsto \{0,1\}^m$ for affine sources with entropy threshold $k \cdot \log q$ and $m = (k-1) \cdot \log q$.*

Outline of the argument. We use our composition techniques to give a different analysis of the construction of [25] which shows that this construction also gives a zero-error disperser. The construction of [25] works by first constructing an affine source extractor D' with small output length $m = \Theta(\log n)$ and then composing it with some function F to obtain an extractor $D(x) = F(x, D'(x))$ that extracts many bits (with rather large error). We observe that the function F designed in [25] is in fact a subsource hitter for affine sources and therefore our composition technique gives that the final construction is a zero-error disperser.

5.1.5 Outline

In Section 5.2 we define the notations used in this chapter. In Section 5.3 we present our main composition theorem. In Section 5.4 we present our results for multiple independent sources. In Section 5.5 we give our results on bit-fixing sources. In Section 5.6 we give our results on affine sources. Finally, in Section 5.7 we give some open problems.

5.2 Preliminaries

In this section we explain the notation used in this chapter. Note that some definitions from the introduction are repeated in more precise form.

General notation: We use $[n]$ to denote the set $\{1, \ldots, n\}$. We use $\mathrm{P}(S)$ to denote the set of subsets of a given set S. Given a string $x \in \{0,1\}^n$ and a set $S \subseteq [n]$ we use x_S to denote the string obtained by restricting x to the indices in S. We denote the length of a string x by $|x|$. Logarithms will always be taken with base 2.

Asymptotic conventions: When stating formal statements in theorems and lemmas, we use the Ω and O signs only to denote *absolute* constants, i.e., not depending on any parameters even if these parameters are considered constants.

Notation for probability distributions: Let Ω be some finite set and let P be a distribution on Ω. (All the probability distributions considered in this chapter are on finite sets). For $B \subseteq \Omega$, we denote the probability of B according to P by $\Pr_P[B]$ or $\Pr[P \subseteq B]$; When $B \in \Omega$, we will also use the notation $\Pr(P = B)$. Given a function $A : \Omega \to U$, we denote by $A(P)$ the distribution induced on U when sampling t by P and calculating $A(t)$. We denote by U_Ω the uniform distribution on Ω. For an integer n, we denote by U_n the uniform distribution on $\{0,1\}^n$. For a distribution P on Ω^d and $j \in [d]$, we denote by P_j the restriction of P to the jth coordinate. We denote by $\mathrm{Supp}(P)$ the support of P. A distribution P is *flat* if it assigns the same probability to all the elements in $\mathrm{Supp}(P)$.

The *statistical distance* between two distributions P and Q on Ω, is defined as

$$\max_{S \subseteq \Omega} |\Pr_P[S] - \Pr_Q[S]| = \frac{1}{2} \sum_{w \in \Omega} |\Pr_P[w] - \Pr_Q[w]| .$$

We say that P is *ϵ-close* to Q if the statistical distance between P and Q is at most ϵ.

Definition 5.9 (conditional distributions). *Let P be a distribution on Ω. Let $C \subseteq \Omega$ be an event such that $\Pr_P(C) > 0$. We define the distribution $(P|C)$ by*

$$\Pr_{(P|C)}[B] = \frac{\Pr_P[B \cap C]}{\Pr_P[C]}$$

for any $B \subseteq \Omega$. Given a function $A : \Omega \to U$, we denote by $(A(P)|C)$ the distribution $A((P|C))$.

We need the notion of a convex combination of distributions.

Definition 5.10 (convex combination of distributions). *Given distributions $P_1, \ldots, P_t$ on a set Ω and coefficients $\mu_1, \ldots, \mu_t \geq 0$ such that $\sum_{i=1}^t \mu_i = 1$, we define the distribution $P \triangleq \sum_{i=1}^t \mu_i \cdot P_i$ by*

$$\Pr_P[B] = \sum_{i=1}^t \mu_i \cdot \Pr_{P_i}[B]$$

for any $B \subseteq \Omega$.

Min-entropy. Recall that the *min-entropy* of a distribution X on Ω is defined as

$$\mathrm{H}_\infty(X) \triangleq \min_{x \in \Omega} \log_2 \frac{1}{\Pr[X = x]}.$$

For a class of distributions $\mathcal{C}$ on Ω, we denote by $\mathcal{C}_k$ the set of distributions in $\mathcal{C}$ that have min-entropy at least k. We need the following standard fact:

Fact 5.2. *Let $k' \geq k$ and let X be a distribution with min-entropy at least k', then, X is a convex combination of flat distributions with min-entropy exactly k.*

We also need the following easy lemma.

Lemma 5.3. *Let X be a distribution on Ω that is ϵ-close to a distribution with min-entropy k. Let $B = \left\{x \in \Omega : \Pr[X = x] \geq 2^{-(k-1)}\right\}$ then $\Pr[X \in B] \leq 2\epsilon$.*

Subsources: We also use the following definition of a subsource.

Definition 5.11. *Let X be a distribution on a set Ω. A distribution X' on Ω is a subsource of X with* measure δ *if $X = \delta \cdot X' + (1-\delta) \cdot X''$ for some $\delta > 0$ and distribution X''. If X' is a subsource of X with measure $\delta \geq 2^{-v} > 0$, we say that X' is a subsource of X with* deficiency v.

We note that this definition is more general than the one considered in [4, 5]. We use it as it is more convenient to do so in this chapter.[3]
We also need the following easy lemma:

Lemma 5.4. *Let X be a distribution on Ω such that $H_\infty(X) \geq k$ and let X' be a subsource of X with deficiency v; then, $H_\infty(X') \geq k - v$.*

Proof. We know that $X = \delta \cdot X' + (1-\delta) \cdot X''$ for some $\delta \geq 2^{-v} > 0$. Thus, for any $x \in \text{Supp}(X')$

$$2^{-k} \geq \Pr[X = x] \geq 2^{-v} \cdot \Pr[X' = x] \Rightarrow \Pr[X' = x] \geq 2^{-(k-v)}.$$

Thus, $\text{H}_\infty(X') \geq k - v$. □

Extractors, dispersers and related objects:

Definition 5.12 (extractors and dispersers)**.** *Let $\mathcal{C}$ be a class of distributions on Ω.*

- *A function $E : \Omega \mapsto \{0,1\}^m$ is an* extractor *for $\mathcal{C}$ with* entropy threshold k *and* error $\epsilon > 0$ *if for every $X \in \mathcal{C}_k$, $E(X)$ is ϵ-close to U_m.*
- *A function $D : \Omega \mapsto \{0,1\}^m$ is a* disperser *for $\mathcal{C}$ with* entropy threshold k *and* error $\epsilon > 0$ *if for every $X \in \mathcal{C}_k$, $|\text{Supp}(D(X))| \geq (1-\epsilon)2^m$.*
- *A disperser D for $\mathcal{C}$ with entropy threshold k is a* zero-error disperser *with* entropy threshold k *if it has error $\epsilon = 0$.*
- *A function $D : \Omega \mapsto \{0,1\}^m$ is a* μ-strongly hitting disperser *for $\mathcal{C}$ with* entropy threshold k *if for every $X \in \mathcal{C}_k$ and for every $z \in \{0,1\}^m$, $\Pr[D(X) = z] \geq \mu$.*

Note that a zero-error disperser is always a μ-strongly hitting disperser for some $\mu > 0$.

We note that all the objects above allow the source X to be a convex combination of distributions in $\mathcal{C}$:

Fact 5.5. *Let $\mathcal{C}$ be a class of distributions on Ω. Let X be a distribution on Ω that is a convex combination of distributions from $\mathcal{C}_k$. Let f be an extractor/disperser/strongly hitting disperser with entropy threshold k. Applying f on X gives the same output guarantee as applying f on distributions in $\mathcal{C}_k$.*

---

[3]The definition in [4, 5] has the additional requirement that there exists a function $f : \Omega \mapsto \{0,1\}$ such that $X' = (X|f(X) = 1)$.

Seeded extractors, dispersers and condensers: We need the notion of seeded extractors. We use the following terminology.

Definition 5.13 (seeded objects).

- *A function* $E : \{0,1\}^n \times \{0,1\}^t \mapsto \{0,1\}^m$ *is a* strong seeded extractor *with* entropy threshold k *and* error ϵ *if for every distribution* X *on* $\{0,1\}^n$ *with* $H_\infty(X) \geq k$, *a* $(1-\epsilon)$ *fraction of* $y \in \{0,1\}^t$ *has that* $E(X,y)$ *is* ϵ*-close to uniform.*
- *A function* $D : \{0,1\}^n \times \{0,1\}^t \mapsto \{0,1\}^m$ *is a* seeded disperser *with* entropy threshold k *and* error ϵ *if for every distribution* X *on* $\{0,1\}^n$ *with* $H_\infty(X) \geq k$,
  $|\{D(x,y) : x \in \mathrm{Supp}(X), y \in \{0,1\}^t\}| \geq (1-\epsilon)2^m$.
- *A function* $C : \{0,1\}^n \times \{0,1\}^t \mapsto \{0,1\}^m$ *is a* strong seeded condenser *with* entropy threshold k, entropy guarantee k' *and* error ϵ *if for every distribution* X *on* $\{0,1\}^n$ *with* $H_\infty(X) \geq k$, *a* $(1-\epsilon)$ *fraction of* $y \in \{0,1\}^t$ *has that* $C(X,y)$ *is* ϵ*-close to some distribution with min-entropy* k'.

5.3 A Composition Theorem

In this section we present a general method for increasing the output length of zero-error dispersers. This is achieved by composing a zero-error disperser with a type of seeded function we call a *subsource hitter*. Our composition applies to both zero-error dispersers and strongly hitting dispersers. We start with the case of zero-error dispersers.

5.3.1 Zero-Error Dispersers

The key component in our composition theorem is the following new object, which we call a "subsource hitter". In the next definition we rephrase Definition 5.4.

Definition 5.14 (subsource hitters). *Let* $\mathcal{C}$ *be a class of distributions on* Ω. *A function* $F : \Omega \times \{0,1\}^t \mapsto \{0,1\}^m$ *is a* subsource hitter *for* $\mathcal{C}$ *with* entropy threshold k *and* subsource entropy $k-v$ *if for every* $X \in \mathcal{C}_k$ *and every* $z \in \{0,1\}^m$ *there exists a* $y \in \{0,1\}^t$ *and a subsource* X' *of* X *such that* $X' \in \mathcal{C}_{k-v}$ *and* $\Pr[F(X',y) = z] = 1$.

The following theorem shows that subsource hitters are tailored to increase the output length of zero-error dispersers.

Theorem 5.15. *Let* $\mathcal{C}$ *be a class of distributions on* Ω.

- *Let* $D' : \Omega \mapsto \{0,1\}^t$ *be a zero-error disperser for* $\mathcal{C}$ *with entropy threshold* $k-v$.

- *Let $F : \Omega \times \{0,1\}^t \mapsto \{0,1\}^m$ be a subsource hitter with entropy threshold k and subsource entropy $k-v$.*

Define $D : \Omega \mapsto \{0,1\}^m$ by $D(x) \triangleq F(x, D'(x))$. Then D is a zero-error disperser for $\mathcal{C}$ with entropy threshold k.

Proof. Let X be a distribution in $\mathcal{C}_k$ and $z \in \{0,1\}^m$. By the guarantee on F we have that there exists a $y \in \{0,1\}^t$ and a subsource X' of X such that $\Pr[F(X',y) = z] = 1$ and $X' \in \mathcal{C}_{k-v}$. Note that X' meets the entropy threshold of D' and therefore there exists $x \in \mathrm{Supp}(X') \subseteq \mathrm{Supp}(X)$ such that $D'(x) = y$. It follows that

$$D(x) = F(x, D'(x)) = F(x, y) = z.$$

□

5.3.2 Strongly Hitting Dispersers

In this section we generalize the composition argument so that it preserves the strongly hitting property. We start by generalizing the notion of subsource hitters:

Definition 5.16 (generalized subsource hitters). *Let $\mathcal{C}$ be a class of distributions on Ω. A function $F : \Omega \times \{0,1\}^t \mapsto \{0,1\}^m$ is a* generalized subsource hitter *for $\mathcal{C}$ with* entropy threshold k, subsource entropy $k-v$, measure α *and* error ϵ *if for every $X \in \mathcal{C}_k$ and $z \in \{0,1\}^m$ at least a $(1-\epsilon)$ fraction of $y \in \{0,1\}^t$ has the property that there exists a subsource X' of X of measure α such that X' is a convex combination of distributions in $\mathcal{C}_{k-v}$ and $\Pr[F(X',y) = z] = 1$.*

The generalized version differs from the original version in two respects:

- We require that there are *many* seeds y that hit z rather than requiring that there exists *one* seed y that hits z.
- We allow X' to be a convex combination of sources in $\mathcal{C}_{k-v}$ rather than requiring that X' itself be in $\mathcal{C}_{k-v}$. This allows X' to have a larger measure in the original source X.

Note that any generalized subsource hitter is a subsource hitter with the same entropy threshold and subsource entropy. (This is because we can replace the subsource X' with one of the components in the convex combination and this component is a subsource of X that meets the requirements of Definition 5.14). The following theorem is analogous to Theorem 5.15 for the case of strongly hitting dispersers.

Theorem 5.17. *Let $\mathcal{C}$ be a class of distributions on Ω.*

- *Let $D' : \Omega \mapsto \{0,1\}^t$ be a μ-strongly hitting disperser for $\mathcal{C}$ with entropy threshold $k-v$.*

- *Let $F : \Omega \times \{0,1\}^t \mapsto \{0,1\}^m$ be a subsource hitter with entropy threshold k, subsource entropy $k - v$, measure α and error ϵ.*

Define $D : \Omega \mapsto \{0,1\}^m$ by $D(x) \triangleq F(x, D'(x))$. Then D is a $((1-\epsilon)2^t\alpha\mu)$-strongly hitting disperser for $\mathcal{C}$ with entropy threshold k.

Before proving the theorem, let us discuss some of the parameters. Note that any μ-strongly hitting disperser with output length m has $\mu \leq 2^{-m}$. Let us suppose that D', which has output length t, comes close to this bound (say that D' is μ-strongly hitting for $\mu = 2^{-t-O(1)}$). If F is also close to optimal in the sense that it has measure close to 2^{-m} (say $\alpha = 2^{-m-O(1)}$), then the "new disperser" D is ν-strongly hitting for $\nu = ((1-\epsilon)2^t\alpha\mu) = 2^{-m-O(1)}$. This means that when composing a "near optimal" strongly hitting disperser using a "near-optimal" generalized subsource hitter we indeed obtain a "near-optimal" strongly hitting disperser with large output length. We now give the proof of the theorem.

Proof. (of Theorem 5.17) We prove the second item. Let X be a distribution in $\mathcal{C}_k$ and $z \in \{0,1\}^m$. By the guarantee on F we have that there exists a set $G \subseteq \{0,1\}^t$ of size $(1-\epsilon)2^t$ such that for every $y \in \{0,1\}^t$ there exists a subsource X'_y of X with measure α such that $\Pr[F(X'_y, y) = z] = 1$ and X'_y is a convex combination of distributions from $\mathcal{C}_{k-v}$. For every $y \in G$ we consider applying D' on X'_y (note that X'_y is a convex combination of distributions in $\mathcal{C}_{k-v}$ which meet the entropy threshold of D'). By Fact 5.5 we have that $\Pr[D'(X'_y) = y] \geq \mu$. Let

$$E_y = \{x : D'(x) = y \wedge F(x,y) = z\}.$$

We can rephrase the former statement and conclude that for every $y \in G$, $\Pr[X'_y \in E_y] \geq \mu$.

Note that for $x \in E_y$ we have that $D(x) = z$. Summing up, we have that:

$$\begin{aligned}
\Pr[D(X) = z] &\geq \textstyle\sum_{y \in G} \Pr[D(X) = z | X \in E_y] \cdot \Pr[X \in E_y] \\
&= \textstyle\sum_{y \in G} \Pr[X \in E_y] \\
&\geq \textstyle\sum_{y \in G} \alpha \cdot \Pr[X'_y \in E_y] \\
&\geq \textstyle\sum_{y \in G} \alpha \cdot \mu \\
&\geq (1-\epsilon) \cdot 2^t \cdot \alpha \cdot \mu.
\end{aligned}$$

□

5.4 Zero-Error Dispersers for Multiple Independent Sources

In this section we apply our composition techniques for the class of "multiple independent sources".

5.4.1 Formal Definition of Multiple Independent Sources

We now give a formal definition of the class of "multiple independent sources". We stress that the terminology used here is slightly different than that used in the introduction (which uses the "standard" terminology in the area). This new terminology is chosen to allow such sources to be handled by theorems that apply to general classes $\mathcal{C}$ (e.g., the composition theorems in Section 5.3).

We consider sources that are composed of ℓ independent high min-entropy distributions. We use the following notation.

Definition 5.18 (ℓ-sources). *A distribution $X = (X_1, \ldots, X_\ell)$ on $\Omega = (\{0,1\}^n)^\ell$ is an ℓ-source if the ℓ distributions $X_1, \ldots, X_\ell$ are independent. An ℓ-source X is a* balanced ℓ-source *if*

$$H_\infty(X_1) = H_\infty(X_2) = \ldots = H_\infty(X_\ell).$$

We say that an ℓ-source X has block entropy *at least k if for every $1 \leq i \leq \ell$, $H_\infty(X_i) \geq k$. We say that an ℓ-source X has* block entropy *exactly k if for every $1 \leq i \leq \ell$, $H_\infty(X_i) = k$.*

Note that a balanced ℓ-source X has min-entropy $k \cdot \ell$ if and only if X has block entropy exactly k. The following lemma is an immediate corollary of Fact 5.2

Lemma 5.6. *Every ℓ-source X with block entropy at least k is a convex combination of ℓ-sources with block entropy exactly k.*

By Fact 5.5 we can restrict our attention to designing dispersers for ℓ-sources with block entropy exactly k (or equivalently to balanced ℓ-sources with min-entropy $\ell \cdot k$), and these dispersers can also be applied on ℓ-sources with block entropy at least k.

5.4.2 A Subsource Hitter for 2-Sources

In this section we construct a subsource hitter for balanced 2-sources. We make use of the "Hadamard extractor" constructed by [70, 14] (see also [19]).

Theorem 5.19. *There exists a constant $c_0 > 0$ such that for sufficently large p there is a poly(p)-time computable extractor $H : (\{0,1\}^p)^2 \mapsto \{0,1\}^m$ for balanced 2-sources with entropy threshold $2 \cdot 0.8p$ and error 2^{-2m} for $m = c_0 p$.*

Our construction of subsource hitters also uses a strong seeded condenser (see Definition 5.13). For different settings of parameters we use different choices of off-the-shelf condensers. We elaborate on these choices later. We now present our construction.

Theorem 5.20. *Let n, k, p be integers such that $n \geq k \geq p \geq 100$. Let c_0 be the constant from Theorem 5.19 and let $m = c_0 p$.*

- *Let $C : \{0,1\}^n \times \{0,1\}^t \to \{0,1\}^p$ be a strong condenser for general sources with entropy threshold k, entropy guarantee $0.9p$ and error $1/100$.*
- *Let $H : (\{0,1\}^p)^2 \to \{0,1\}^m$ be the 2-source extractor from Theorem 5.19. (This extractor has entropy threshold $2 \cdot 0.8p$ and error 2^{-2m}).*

Define the function $F : (\{0,1\}^n)^2 \times \{0,1\}^t \to \{0,1\}^m$ by $F(x,y) = H(C(x_1,y), C(x_2,y))$; then,

- *F is a subsource hitter for balanced 2-sources with entropy threshold $2 \cdot k$ and subsource entropy $2 \cdot (k - 3p)$.*
- *F is a generalized subsource hitter for balanced 2-sources with entropy threshold $2 \cdot k$, subsource entropy $2 \cdot (k - 3p)$, measure $2^{-(m+1)}$ and error $1/10$.*

Proof. Let X be a balanced 2-source on $(\{0,1\}^n)^2$ with min-entropy at least $2k$. Note that this means that X_1, X_2 are independent distributions with min-entropy k. We have that C is a strong condenser with this entropy threshold and therefore for any distribution V with min-entropy k a $(1-1/100)$ fraction of $y \in \{0,1\}^t$ is good in the sense that $C(V,y)$ is $(1/100)$-close to having min-entropy $0.9p$. By a union bound it follows that a $1 - 2/100$ fraction of $y \in \{0,1\}^t$ satisfies this property for both X_1, X_2 simultaneously, namely that: both $C(X_1,y)$ and $C(X_2,y)$ are $(1/100)$-close to having min-entropy $0.9p$. We call such $y \in \{0,1\}^t$ "good seeds". Fix some good seed y and let $R_1 = C(X_1,y)$ and $R_2 = C(X_2,y)$. We define

$$B_1' = \left\{ r \in \{0,1\}^p : \Pr[R_1 = r] < 2^{-(p+10)} \right\}.$$

Note that

$$\Pr[R_1 \in B_1'] \leq \sum_{r \in B_1'} \Pr[R_1 = r] \leq 2^p \cdot 2^{-(p+10)} \leq 2^{-10}.$$

We define

$$B_1'' = \left\{ r \in \{0,1\}^p : \Pr[R_1 = r] > 2^{-(0.9p-1)} \right\}.$$

By Lemma 5.3 we have that $\Pr[R_1 \in B_1''] \leq 2/100$. Let $B_1 = B_1' \cup B_1''$ and note that $\Pr[R_1 \in B_1] \leq 2/100 + 2^{-10} \leq 1/10$.

We can repeat the same argument for R_2 and define subsets B_2, B_2', B_2'' in an analogous way and conclude that $\Pr[R_2 \in B_2] \le 1/10$. Let us consider the events $E_1 = \{R_1 \notin B_1\}$, $E_2 = \{R_2 \notin B_2\}$ and $E = E_1 \cap E_2$ (note that we think about these events as events over the original distribution X). Let $V = (X|E)$. Note that $V_1 \sim (X_1|E_1)$ and $V_2 \sim (X_2|E_2)$ and that the two distributions V_1, V_2 are independent. Let us estimate the min-entropy of the distribution $C(V_1, y)$: For any r in the support of $C(V_1, y)$ we have that

$$\Pr[C(V_1,y)=r] = \Pr[C(X_1,y)=r|E_1] = \Pr[R_1=r|E_1]$$
$$\le \frac{\Pr[R_1=r]}{\Pr[E_1]} \le \frac{2^{-(0.9p-1)}}{9/10} \le 2^{-(0.9p-2)}.$$

Thus, we conclude that $C(V_1, y)$ has min-entropy at least $0.9p - 2 \ge 0.8p$. We can use the same argument for $C(V_2, y)$. We have that the two distributions $C(V_1, y), C(V_2, y)$ are independent and meet the entropy threshold of the extractor H. We conclude that $H(C(V_1, y), C(V_2, y))$ is 2^{-2m}-close to uniform. Fix some string $z \in \{0,1\}^m$. It follows that

$$\Pr[H(C(V_1,y),C(V_2,y)) = z] \ge 2^{-m} - 2^{-2m}.$$

It follows that

$$\begin{aligned}
\Pr[E \wedge H(R_1,R_2)=z] &= \Pr[E]\cdot\Pr[H(R_1,R_2)=z|E] \\
&= \Pr[E_1]\cdot\Pr[E_2]\cdot\Pr[H(C(V_1,y),C(V_2,y))=z] \\
&\ge (9/10)^2\cdot(2^{-m}-2^{-2m}) \\
&\ge 2^{-(m+1)}.
\end{aligned}$$

We say that a pair $(r_1, r_2) \in (\{0,1\}^p)^2$ is *useful* (with respect to a good seed $y \in \{0,1\}^t$ and a $z \in \{0,1\}^m$) if $r_1 \notin B_1$, $r_2 \notin B_2$ and $H(r_1, r_2) = z$. Summing up what we did so far we have that a $(1 - 2/100)$ fraction of $y \in \{0,1\}^t$ are good seeds, and for any such good seed $y \in \{0,1\}^t$ and $z \in \{0,1\}^m$ we have that with probability $2^{-(m+1)}$ the pair $(C(X_1, y), C(X_2, y))$ is useful. For any useful pair (r_1, r_2) we define a subsource $X^{(r_1,r_2)}$ of X by

$$X^{(r_1,r_2)} = (X|C(X_1,y)=r_1 \wedge C(X_2,y)=r_2).$$

We claim that:

Claim 5.0.2. *For every $(r_1, r_2) \in (\{0,1\}^p)^2$ useful with respect to a good seed y and $z \in \{0,1\}^m$ we have that*

- $\Pr[F(X^{(r_1,r_2)}, y) = z] = 1$.
- *$X^{(r_1,r_2)}$ is a convex combination of balanced 2-sources with min-entropy exactly $2 \cdot (k - 3p)$.*

Proof. (of Claim 5.0.2) The first item follows because for every $x \in \text{Supp}(X^{(r_1,r_2)})$ we have that

$$F(x,y) = H(C(x_1,y),C(x_2,y)) = H(r_1,r_2) = z.$$

For the second item, note that the two distributions $X_1^{(r_1,r_2)}$ and $X_2^{(r_1,r_2)}$ are independent. Furthermore:

$$\Pr[C(X_1,y){=}r_1 \wedge C(X_2,y){=}r_2] {=} \Pr[C(X_1,y) = r_1]{\cdot}\Pr[C(X_2,y){=}r_2] {\geq} (2^{-(p+10)})^2 {\geq} 2^{-3p}$$

It follows that $X^{(r_1,r_2)}$ is a deficiency $3p$ subsource of X. By Lemma 5.4 we have that $\text{H}_\infty\left(X^{(r_1,r_2)}\right) \geq 2k - 3p$. It follows that $X^{(r_1,r_2)}$ has block entropy at least $k-3p$ and by Lemma 5.6 it is a convex combination of balanced 2-sources with block entropy exactly $k - 3p$ (or equivalently balanced 2-sources with min-entropy $2 \cdot (k - 3p)$). □

We are now ready to prove Theorem 5.20.

Let us first prove the first item, which says that F is a subsource hitter. Fix some good seed $y \in \{0,1\}^t$ and $z \in \{0,1\}^m$. Let (r_1, r_2) be a useful pair with respect to y and z. By the first item of Claim 5.0.2 we have that $X^{(r_1,r_2)}$ is a convex combination of balanced 2-source with min-entropy exactly $2{\cdot}(k-3p)$. Let X' be one of the components in this convex combination that appears with a positive coefficient. We have that X' is a subsource of $X^{(r_1,r_2)}$ which is in turn a subsource of X. Furthermore, by the second item of Claim 5.0.2 and as $\text{Supp}(X') \subseteq \text{Supp}(X^{(r_1,r_2)})$ we have that $\Pr[F(X',y) = z] = 1$.

We now prove the second item, that is, that F is a generalized subsource hitter. We have that a $(1 - 1/10)$ fraction of $y \in \{0,1\}^t$ are good seeds. Fix some good seed y and $z \in \{0,1\}^m$. We define

$$X' = (X|(C(X_1,y),C(X_2,y)) \text{ are a useful pair}).$$

We have already seen before that that X' has measure $2^{-(m+1)}$ as a subsource of X. Furthermore, X' is a convex combination of the sources $X^{(r_1,r_2)}$ for useful pairs (r_1,r_2). By Claim 5.0.2 each one of the latter sources is a convex combination of balanced 2-sources with min-entropy $2{\cdot}(k-3p)$. Thus, overall X' is a convex combination of balanced 2-sources with min-entropy $2 \cdot (k - 3p)$. For every $x \in \text{Supp}(X')$ there exists a useful pair (r_1,r_2) such that $x \in \text{Supp}(X^{r_1,r_2})$, and we already showed that for such x we have that $F(x,y) = z$. □

5.4.3 Zero-Error Dispersers for 2-Sources

We now plug in specific choices of strong seeded condensers to obtain specific results.

High entropy threshold

Our first choice is a condenser by Raz [53]. This condenser has the advantage that it has a constant-length seed. However, it only works when the entropy threshold is a constant fraction of the length.

Theorem 5.21. *[53] For every $\delta > 0$ there is a $\beta > 0$ and integer t such that for sufficiently large n there is a poly(n)-time computable strong seeded condenser $C : \{0,1\}^n \times \{0,1\}^t \mapsto \{0,1\}^p$ with $p = \beta n$ entropy threshold δn, entropy guarantee $0.9p$ and error $1/100$.*

Plugging Theorem 5.21 into Theorem 5.20 we obtain the following Corollary.

Corollary 5.1. *For every $\delta > 0$ there is an $\eta > 0$ and an integer t such that for sufficiently large n and $m = \eta n$:*

- *There is a poly(n)-time computable generalized subsource hitter F : $(\{0,1\}^n)^2 \times \{0,1\}^t \mapsto \{0,1\}^m$ for balanced 2-sources with entropy threshold $2 \cdot \delta n$, subsource entropy δn, measure $2^{-(m+1)}$ and error $1/10$.*
- *Any poly(n)-time computable μ-strongly hitting disperser $D' : (\{0,1\}^n)^2 \mapsto \{0,1\}^t$ for balanced 2-sources with entropy threshold δn can be transformed into a poly(n)-time computable $(\mu 2^{t-m-2})$-strongly hitting disperser $D : (\{0,1\}^n)^2 \mapsto \{0,1\}^m$ for balanced 2-sources with entropy threshold $2\delta n$.*

We can apply the second item in the corollary above on the strongly hitting disperser of Barak et al. [4].

Theorem 5.22 ([4])**.** *For every $\delta > 0$ and integer t there exists a $\mu > 0$ such that for sufficiently large n there is a poly(n)-time computable μ-strongly hitting disperser $D : (\{0,1\}^n)^2 \mapsto \{0,1\}^t$ with entropy threshold δn.*

Applying the aforementioned transformation we get that:

Theorem 5.23. *For every $\delta > 0$ there exist a $\nu > 0$ and an $\eta > 0$ such that for sufficiently large n there is a poly(n)-time computable $(\nu 2^{-m})$-strongly hitting disperser $D : (\{0,1\}^n)^2 \mapsto \{0,1\}^m$ with entropy threshold δn and $m = \eta n$.*

Arbitrary entropy threshold

In order to handle lower entropy thresholds we use a strong seeded extractor (which is in particular a strong seeded condenser).

Theorem 5.24 ([42, 34])**.** *There exists a number c such that for every sufficiently large k, n there is a poly(n)-time computable strong seeded extractor $E : \{0,1\}^n \times \{0,1\}^{c \log n} \mapsto \{0,1\}^m$ for entropy threshold k, error $1/100$ and $m = k/2$.*

Plugging Theorem 5.24 into Theorem 5.20 we obtain the following corollary.

Corollary 5.2. *There exist $\eta > 0$ and c such that for every sufficiently large k, n and $m = \eta k$:*

- *There is a poly(n)-time computable generalized subsource hitter $F : (\{0,1\}^n)^2 \times \{0,1\}^{t=c\log n} \mapsto \{0,1\}^m$ for balanced 2-sources with entropy threshold $2 \cdot k$, subsource entropy k, measure $2^{-(m+1)}$ and error $1/10$.*
- *Any poly(n)-time computable μ-strongly hitting disperser $D' : (\{0,1\}^n)^2 \mapsto \{0,1\}^{c\log n}$ for balanced 2-sources with entropy threshold k can be transformed into a poly(n)-time computable $(\mu 2^{t-m-2})$-strongly hitting disperser $D : (\{0,1\}^n)^2 \mapsto \{0,1\}^m$ for balanced 2-sources with entropy threshold $2 \cdot k$.*

Barak et al. [5] construct zero-error dispersers for entropy threshold $k = n^{o(1)}$. One can hope to apply Corollary 5.2 to increase the output length of these dispersers. However, the construction of [5] only achieves output length $m = 1$. We note that by Corollary 5.2 improving the output length to $m = c\log n$ will immediately give further improvement to $m = \Omega(k)$.

5.4.4 Zero-Error Dispersers for $O(1)$-Sources

In the previous section we constructed zero-error dispersers for balanced 2-sources with entropy threshold $k = \delta n$ for any constant $\delta > 0$. We now give constructions that have the disadvantage that they require $\ell > 2$ sources for $\ell = O(1)$. However, they achieve lower entropy thresholds.

We use an ℓ-source extractor constructed by Rao [50]. The version we use here has better analysis that provides low error and is due to Barak et al. [5].

Theorem 5.25 ([50, 5]). *There is a $\gamma > 0$ such that for every sufficiently large $k \leq n$ there are integers $\ell = O(\frac{\log n}{\log k})$, $m = k^\gamma$ and a poly(n)-time computable extractor $E : (\{0,1\}^n)^\ell \mapsto \{0,1\}^m$ for balanced ℓ-sources with entropy threshold $\ell \cdot k$ and error $\epsilon < 2^{-m+1}$.*

Note that by Fact 5.1 such an extractor is in particular a μ-strongly hitting disperser for $\mu = 2^{-(m+1)}$. We now show how to improve the output length to $m = \Omega(k)$ while preserving this property.

Theorem 5.26. *There are numbers $c', \eta > 0$ such that for every sufficiently large k, n such that $k \geq (\log n)^{c'}$ there are integers $\ell = O(\frac{\log n}{\log k})$, $m = \eta k$ and a poly(n)-time computable $2^{-(m+3)}$-strongly hitting disperser $D : (\{0,1\}^n)^\ell \mapsto \{0,1\}^m$ for balanced ℓ-sources with entropy threshold $\ell \cdot k$.*

Proof. By Corollary 5.2 there exist $\eta > 0$ and c such that for sufficiently large $k \leq n$ and $m = \eta k$ there is a poly(n)-time computable generalized subsource hitter $F : (\{0,1\}^n)^2 \times \{0,1\}^{c\log n} \mapsto \{0,1\}^m$ for balanced 2-sources with entropy threshold $2 \cdot k$, subsource entropy k, measure $2^{-(m+1)}$ and error $1/10$. Let $t = c \log n$.

Let E be the extractor from Theorem 5.25 (for the same k, n) and let γ, ℓ, m be the parameters associated with it. The extractor E has output length k^γ; by choosing c' to be a sufficiently large constant as a function of the constants c, γ we have that $k \geq (\log n)^{c'}$ and so $k^\gamma \geq c \log n$. We can thus chop the output of E to length $t = c \log n$. Note that E is a $2^{-(t+1)}$-strongly hitting disperser. Let $\ell' = \ell + 2$. We construct a zero-error disperser D for balanced ℓ'-sources with entropy threshold $\ell' \cdot k$ by

$$D(x_1, \dots, x_{\ell'}) = F(x_{\ell+1}, x_{\ell+2}, E(x_1, \dots, x_\ell)).$$

Indeed, let $X = (X_1, \dots, X_{\ell'})$ be a balanced ℓ'-source with min-entropy at least $\ell' \cdot k$. We consider the balanced 2-source $(X_{\ell+1}, X_{\ell+2})$. By the properties of F we have that for every $z \in \{0,1\}^m$ a $9/10$ fraction of $y \in \{0,1\}^t$ (which we call good seeds) has

$$\Pr[F(X_{\ell+1}, X_{\ell+2}, y) = z] \geq 2^{-(m+1)}. \tag{5.1}$$

(Note that here we're not even using the property that F hits z on a well-structured subsource. We're only using the fact that F hits z with positive probability.) We also consider the balanced ℓ-source $(X_1, \dots, X_\ell)$. As E is a $2^{-(t+1)}$-strongly hitting disperser we have that for every $y \in \{0,1\}^t$

$$\Pr[E(X_1, \dots, X_\ell) = y] > 2^{-(t+1)}. \tag{5.2}$$

For every good seed $y \in \{0,1\}^t$ we have that if the two events in (5.1) and (5.2) are independent, and therefore the probability that they occur simultaneously is at least $2^{-(t+1)} \cdot 2^{-(m+1)}$. Whenever this happens we have that $D(X_1, \dots, X_{\ell'}) = z$. Summing up over the $\frac{9}{10} \cdot 2^t$ good seeds y we have that

$$\Pr[D(X_1, \dots, X_{\ell'}) = z] \geq \frac{9}{10} \cdot 2^t \cdot 2^{-(t+1)} \cdot 2^{-(m+1)} \geq 2^{-(m+3)}.$$

□

5.4.5 Rainbows and Implicit $O(1)$ Probe Search

In this section we discuss an application of zero-error dispersers to the problem of *implicit probe search*. Loosely speaking, this is the problem of searching for an element in a table with few probes, when no additional information other than the elements themselves is stored.

Definition 5.27 (implicit probe search scheme). *For integer parameters n, k, q, the implicit probe search problem is as follows: Store a subset $S \subseteq \{0,1\}^n$ of size 2^k in a table T of size 2^k (where every table entry holds only a single element of S) such that given $x \in \{0,1\}^n$ we can determine whether $x \in S$ using q queries on T. A solution to this problem is called an* implicit q-probe scheme with table size 2^k and domain size 2^n.

Fiat and Naor [24] investigated implicit $O(1)$-probe schemes, i.e., schemes where the number of queries is a constant not depending on n and k. They showed that this problem is unsolvable when n is large enough relative to k (this improves a previous bound by Yao [75]). They also gave an efficient implicit $O(1)$-probe scheme whenever $k = \delta \cdot n$ for any constant $\delta > 0$. They did this by reducing the problem to the task of constructing a combinatorial object called a *rainbow*.

Definition 5.28. *[24]*

- *A* t-sequence *over a set* U *is a sequence of length* t*, without repetitions, of elements in* U*.*
- *An* (n, k, t)*-rainbow is a coloring of all* t*-sequences over* $\{0,1\}^n$ *with* 2^k *colors such that for any* $S \subseteq \{0,1\}^n$ *of size* 2^k*, the* t*-sequences over* S *are colored in all colors.*

Fiat and Naor showed that rainbows imply implicit probe schemes.

Theorem 5.29. *[24] Fix any integers* n, k *with* $\log n \leq k \leq n$*. Given a poly*(n)*-time computable* (n, k, t)*-rainbow we can construct a poly*(n)*-time computable implicit* $O(t)$*-probe scheme with table size* 2^k *and domain size* 2^n*. In particular, when* t *is constant we get an implicit* $O(1)$*-probe scheme.*

The following theorem shows that we can construct rainbows from zero-error dispersers for multiple independent sources.

Theorem 5.30.[4] *Let* $0 < \eta < 1$ *be any constant, and let* n, k *and* t *be integers with* $\log n \leq k \leq n$*. Let* $G : \{0,1\}^{t \cdot n} \mapsto \{0,1\}^m$ *be a* $2 \cdot t^2/2^k$*-strongly hitting disperser for balanced* t*-sources with entropy threshold* $t \cdot k$*, where* $m = \eta \cdot k$*. Let* $l = \lceil 1/\eta \rceil$*, and define* $\bar{G} : (\{0,1\}^{t \cdot n})^l$ *by*

$$\bar{G}(X_1, \ldots, X_l) \triangleq G(X_1) \circ \ldots \circ G(X_l).$$

Then $\bar{G}$ *is an* $(n, k, t \cdot l)$*-rainbow.*

Proof. Fix a subset $S \subseteq \{0,1\}^n$ with $|S| = 2^k$. Let X be the uniform distribution on $S \subseteq \{0,1\}^n$. Thus X has min-entropy k. Let X^{*t} denote the distribution made of t independent copies of X. X^{*t} is a t-source with block entropy exactly k. Therefore for any $z \in \{0,1\}^m$,

$$\Pr[G(X^{*t}) = z] \geq 2 \cdot t^2/2^k.$$

On the other hand, the probability that X^{*t} outputs a string $(x_1, \ldots, x_t), x_i \in \{0,1\}^n$ with $x_i = x_j$ for some $1 \leq i \neq j \leq t$ is at most $t^2/2^k < 2 \cdot t^2/2^k$. Therefore there must be $(x_1, \ldots, x_t) \in \text{Supp}(X^{*t})$ 'without repetitions', i.e.,

[4] An error in the proof of this Theorem was recently discovered. See the paper 'Increasing the Output Length of Zero-Error Dispersers' on the author's website https://sites.google.com/site/arielgabizon1/ for a corrected version.

$x_i \neq x_j$ for all $i \neq j$, such that $G(x_1, \ldots, x_t) = z$. Note that this exactly means there is a t-sequence of elements of S that G 'colors' z. Concatenating l copies of G we get an output of length $\lceil 1/\eta \rceil \cdot \eta k \geq k$ which gives us 2^k colors (of course we can truncate the output if it is larger than k). This shows that $\bar{G}$ is an $(n, k, t \cdot l)$-rainbow. □

Plugging in our strongly hitting disperser for multiple independent sources we get the following implicit probe scheme.

Corollary 5.3. *Fix any constant* $0 < \delta < 1$. *For every sufficiently large* n *and* $k = n^\delta$ *there is a poly(n)-time computable implicit* $O(1/\delta)$*-probe scheme with table size* 2^k *and domain size* 2^n.

Proof. Fix k and n with $k = n^\delta$ for some constant $\delta > 0$. Using Theorem 5.26, we get a 2^{-m+3}-strongly hitting disperser $D : (\{0,1\}^n)^\ell \mapsto \{0,1\}^m$ for balanced ℓ-sources with entropy threshold $k \cdot \ell$ where $\ell = O(\frac{\log n}{\log k}) = O(1/\delta)$ and $m = \eta \cdot k$ for some constant $0 < \eta < 1$ (not depending on δ). Applying Theorem 5.30 we get an $(n, k, O(1/\delta))$-rainbow, and therefore, by Theorem 5.29, an implicit $O(1/\delta)$-probe scheme with table size 2^k and domain size 2^n. □

5.5 Zero-Error Dispersers for Bit-Fixing Sources

In this section we construct dispersers for the family of *bit-fixing sources* introduced by Chor et al. [15]. A distribution X over $\{0,1\}^n$ is a bit-fixing source if there is a subset $S \subseteq [n]$ of "good indices" such that the bits X_i for $i \in S$ are independent fair coins and the rest of the bits are fixed.

Definition 5.31 (bit-fixing sources). *A distribution* X *over* $\{0,1\}^n$ *is an* (n,k)-bit-fixing source *if there exists a subset* $S = \{i_1, \ldots, i_k\} \subseteq [n]$ *such that* $X_{i_1}, X_{i_2}, \ldots, X_{i_k}$ *is uniformly distributed over* $\{0,1\}^k$ *and for every* $i \notin S$, X_i *is constant. The class of* bit-fixing sources over $\{0,1\}^n$ *is the class of all* (n,k)*-bit-fixing sources for some* $1 \leq k \leq n$.

Our construction of zero-error dispersers for bit-fixing sources works by reducing to the case of independent sources. More specifically, we show that a subsource hitter for independent sources implies a subsource hitter for bit-fixing sources. This is done by sampling two blocks from the source such that each block contains a linear fraction of the entropy, and so does the remaining part of the source.

We will do this using the following partitioning lemma based on pairwise independence.

Lemma 5.7. *For any integers* k *and* n *with* $64 < k \leq n$, *there is a poly(n)-time computable function* $P : \{0,1\}^{2\log n} \mapsto (\mathrm{P}([n]))^4$ *returning a partition* *of $[n]$ into four disjoint sets $P(y)_1 \cup P(y)_2 \cup P(y)_3 \cup P(y)_4 = [n]$ such that for any (n,k)-bit-fixing source X, there exists a $y \in \{0,1\}^{2\log n}$ such that for every $i \in [4]$, $X_{P(y)_i}$ is an (n',k')-bit-fixing source for some $n' \leq n$ and $k' \geq k/8$.*

Proof. We use y as a random seed to generate pairwise independent variables $Z_1,\ldots,Z_n \in [4]$ (there are constructions which use $2\log n$ bits to generate such variables [13]). For $i = 1,\ldots,4$ define the subset $P(y)_i \subseteq [n]$ by $P(y)_i \triangleq \{j : Z_j = i\}$. Assume without loss of generality that the 'good indices' of X are $\{1,\ldots,k\}$. Fix any $i \in [4]$. For $j \in [n]$ define the random variable X_j by $X_j = 1$ if $Z_j = i$, and 0 otherwise. Then, for $j \in [n]$, $E(X_j) = 1/4$ and $Var(X_j) \leq 1/4$. Furthermore, for $j \neq l$ X_j and X_l are independent and $cov(X_j, X_l) = 0$. Define $Y = \sum_{j=1}^{k} X_j$. We have $E(Y) = k/4$ and $Var(Y) = \sum_{j=1}^{k} Var(X_j) \leq k/4$. Therefore, by Chebychev's inequality

$$\Pr[|Y - k/4| \geq k/8] \leq k/4 \cdot (8/k)^2 \leq 16/k.$$

Note that Y is exactly the number of good indices in $P(Y)_i$. Thus, using the union bound, with probability $1 - 64/k$ over y, for every $i \in [4]$, $P(Y)_i$ contains at least $k/8$ good indices of X. In particular, when $k > 64$ there exists a y such that for every $i \in [4]$, $X_{P(y)_i}$ is an (n',k')-bit-fixing source for some $n' \leq n$ and $k' \geq k/8$. □

Using the above lemma, we show how to construct a subsource hitter for bit-fixing sources from a subsource hitter for 2-sources.

Lemma 5.8. *Fix any integers k and n with $64 < k \leq n$.*

- *Let $G : (\{0,1\}^n \times \{0,1\}^n) \times \{0,1\}^t \mapsto \{0,1\}^m$ be a subsource hitter for balanced 2-sources with entropy threshold $2 \cdot k/8$ and subsource entropy l.*
- *Let $P : \{0,1\}^{2\log n} \mapsto (\mathrm{P}([n]))^4$ be the partitioning function from Lemma 5.7.*

Define the function $F : \{0,1\}^n \times \{0,1\}^{2\log n + t} \mapsto \{0,1\}^m$ by

$$F(x,(y,y')) \triangleq G((x_{P(y)_1}, x_{P(y)_2}), y')$$

(we pad $x_{P(y)_1}$ and $x_{P(y)_2}$ with zeros to make them n-bit strings). Then F is a subsource hitter for bit-fixing sources with entropy threshold k and subsource entropy $k/4$.

Proof. Let X be an (n,k)-bit-fixing source. Using Lemma 5.7, we can fix a $y \in \{0,1\}^{2\log n}$ such that for every $i \in [4]$, $X_i \triangleq X_{P(y)_i}$ is an (n',k')-bit-fixing source for some $n' \leq n$ and $k' \geq k/8$. Note that (X_1,X_2) is a 2-source with block entropy at least $k/8$. Fix any $z \in \{0,1\}^m$, and fix $y' \in \{0,1\}^t$ such that there is a subsource (X_1', X_2') of (X_1,X_2) with min-entropy at least l such that $\Pr[G((X_1',X_2'),y') = z] = 1$. (Such a subsource exists, as by Lemma

5.6, (X_1, X_2) has a subsource (X_1^*, X_2^*), which is a balanced 2-source with entropy threshold $2 \cdot k/8$. (X_1^*, X_2^*) contains such a subsource (X_1', X_2') by the guarantee of G and (X_1', X_2') is also a subsource of (X_1, X_2)). Fix an arbitrary $x' \in \mathrm{Supp}(X_1', X_2')$ and let $X' \triangleq (X|(X_1, X_2) = x')$. Note that X' is an (n, k')-bit-fixing source for some $k' \geq k/4$ as $P(y)_3 \cup P(y)_4$ contains at least $k/4$ good indices. For any $x \in X'$, we have

$$F(x, (y, y')) = G(x', y') = z$$

and thus $\Pr[F(X', (y, y')) = z] = 1$. As X' is a subsource of X with min-entropy at least $k/4$ this proves that F is a subsource hitter for bit-fixing sources with entropy threshold k and subsource entropy $k/4$. □

Plugging in the subsource hitter for 2-sources from Corollary 5.2 we get the following.

Corollary 5.4. *There exist constants $c > 0$ and $0 < \eta < 1$ such that for every sufficiently large $k \leq n$ there is a poly(n)-time computable subsource hitter $F : \{0,1\}^n \times \{0,1\}^{c \log n} \mapsto \{0,1\}^m$ for bit-fixing sources with entropy threshold k and subsource entropy $k/4$, where $m = \eta \cdot k$.*

We will use as a component the following result of Rao [52].

Theorem 5.32. *There exist constants $c > 0$ and $0 < d < 1$ such that for every $k \leq n$ with $k \geq \log^c n$, there is a poly(n)-time computable zero-error disperser $D : \{0,1\}^n \mapsto \{0,1\}^t$ for bit-fixing sources with entropy threshold k, where $t = k^d$.*

Remark 5.1. *Actually, the result above of [52] applies to 'low-weight affine sources', which are much more general than bit-fixing sources.*

We can now prove our main result for bit-fixing sources.

Theorem 5.33. *There exist constants $c > 0$ and $0 < \eta < 1$ such that for every sufficiently large $k \leq n$ with $k \geq \log^c n$ there is a poly(n)-time computable zero-error disperser $D : \{0,1\}^n \mapsto \{0,1\}^m$ for bit-fixing sources with entropy threshold k, where $m = \eta \cdot k$.*

Proof. Using Theorem 5.32 and Corollary 5.4 we can choose a large enough constant c' such that for some constants $0 < d, \eta < 1$, for any $k \geq \log^{c'} n$, we have the following explicit components:

- A zero-error disperser $D' : \{0,1\}^n \mapsto \{0,1\}^{(k/4)^d}$ for bit-fixing sources with entropy threshold $k/4$.
- A subsource hitter $F : \{0,1\}^n \times \{0,1\}^{c' \log n} \mapsto \{0,1\}^{\eta \cdot k}$ for bit-fixing sources with entropy threshold k and subsource entropy $k/4$.

To use Theorem 5.15, we need to make sure D''s output is as long as F's seed. Assuming $k \geq \log^{2/d} n$ we have

$$(k/4)^d \geq (\log^{2/d} n/4)^d \geq (\log^2 n)/4^d \geq c' \cdot \log n$$

for large enough n. Thus, using Theorem 5.15, we get a zero-error disperser $D : \{0,1\}^n \mapsto \{0,1\}^{\eta \cdot k}$ for bit-fixing sources with entropy threshold k. Taking $c = \max\{c', 2/d\}$ we are done. □

5.6 Zero-Error Dispersers for Affine Sources

Denote by $\mathbb{F}_q$ the finite field with q elements. Denote by $\mathbb{F}_q^n$ the n-dimensional vector space over $\mathbb{F}_q$.

We formally define *affine sources.*

Definition 5.34 (affine source). *A distribution X over $\mathbb{F}_q^n$ is an $(n,k)_q$-*affine source *if it is uniformly distributed over an affine subspace of dimension k. That is, X is sampled by choosing $t_1, \ldots, t_k$ uniformly and independently in $\mathbb{F}_q$ and calculating*

$$\sum_{j=1}^{k} t_j \cdot a^{(j)} + b$$

for some $a^{(1)}, \ldots, a^{(k)}, b \in \mathbb{F}_q^n$ such that $a^{(1)}, \ldots, a^{(k)}$ are linearly independent. The class of affine sources over $\mathbb{F}_q^n$ *is the class of all $(n,k)_q$-affine sources for some $1 \leq k \leq n$.*

Note that an $(n,k)_q$-affine source has min-entropy $k \cdot \log q$. We will use affine source *extractors* (though we just use the fact that they are zero-error dispersers when their error is small enough), which we now formally define.

Definition 5.35 (deterministic affine source extractor). *A function $D : \mathbb{F}_q^n \rightarrow \{0,1\}^t$ is a deterministic (k,ϵ)-affine source extractor if for every $(n,k)_q$-affine source X the distribution $D(X)$ is ϵ-close to uniform.*

In our construction of a zero-error disperser for affine sources, we use two components from [25]. The first is an extractor for $(n,1)_q$-affine sources with error exponentially small in the output length. The second is a small set of linear mappings $T_u : \mathbb{F}_q^n \rightarrow \mathbb{F}_q^k$ such that for any affine subspace $A \subseteq \mathbb{F}_q^n$ of dimension k, one of (actually, most of) the mappings in the set maps A *onto* $\mathbb{F}_q^k$.

Remark 5.2. *Our construction will turn out to be identical to the affine source extractor construction of [25]. Thus we prove that the extractor of [25], while having large error in relation to the output length, still outputs all elements with positive probability.*

We will use the following instantiation of a lemma from [25].

Lemma 5.9 (Lemma 5.5 from [25] with $\delta = 4/5$). *For every sufficiently large prime power q and integer n with $q \geq n^9$, there is a poly(n)-time computable deterministic $(1, \epsilon)$-affine source extractor $D : \mathbb{F}_q^n \to \{0,1\}^t$ where $\epsilon \leq q^{-1/5}$ and $t = \lceil (1/6) \log q \rceil$.*

Since the error in the above lemma is exponentially small in the output length, D is a also a zero-error disperser.

Corollary 5.5. *Fix any sufficiently large prime power q and any integer n such that $q \geq n^9$. There is a poly(n)-time computable zero-error disperser $D : \mathbb{F}_q^n \mapsto \{0,1\}^t$ for affine sources with entropy threshold $\log q$, where $t = \lceil (1/6) \log q \rceil$.*

Proof. $D : \mathbb{F}_q^n \to \{0,1\}^t$ will simply be the function from Lemma 5.9. Fix an $(n,1)_q$-affine source X and an element $z \in \{0,1\}^t$. From Lemma 5.9, we know that

$$\Pr[D(X) = z] \geq \frac{1}{2^t} - q^{-1/5} \geq \frac{1}{2} \cdot q^{-1/6} - q^{-1/5} > 0$$

(for large enough q). □

Our subsource hitter will be based on the following construction of a set of matrices from [25]. Given $u \in \mathbb{F}_q$ and an integer k, we define a $k \times n$ matrix $T_{u,k}$ by $(T_{u,k})_{j,i} = u^{ji}$ (where ji is an integer product). That is,

$$T_{u,k}(x) = \left(\sum_{i=1}^{n} x_i \cdot u^i, \sum_{i=1}^{n} x_i \cdot u^{2i}, \ldots, \sum_{i=1}^{n} x_i \cdot u^{ki} \right)$$

for $x \in \mathbb{F}_q^n$.

Lemma 5.10 (Lemma 6.1 in [25]). *Fix any field $\mathbb{F}_q$ and integers n, k such that $q \geq n \cdot k^2$. Fix any affine subspace $A \subseteq \mathbb{F}_q^n$ of dimension at least k. There are at most $n \cdot k^2$ elements $u \in \mathbb{F}_q$ such that $T_{u,k}(A) \subsetneq \mathbb{F}_q^k$.*

Corollary 5.6. *Fix any sufficiently large prime power q and any integers n, k such that $q \geq n^{18}$ and $2 \leq k < n$. Let $s = 2^t$ where $t = \lceil (1/6) \log q \rceil$. Let $U = \{u_1, \ldots, u_s\}$ be a set of distinct elements in $\mathbb{F}_q$. We identify U with $\{0,1\}^t$. The function $F : \mathbb{F}_q^n \times \{0,1\}^t \to \mathbb{F}_q^{k-1}$ defined by*

$$F(x, u) \triangleq T_{u,k-1}(x)$$

is a subsource hitter for affine sources with entropy threshold $k \cdot \log q$ and deficiency $\log q$.

Proof. X is uniformly distributed on an affine subspace A of dimension k, i.e., $\mathrm{Supp}(X) = A$. Since $|U| = s \geq q^{1/6} > n \cdot (k-1)^2$, by Lemma 5.10 there is $u \in U$ such that $T_{u,k-1}(A) = \mathbb{F}_q^{k-1}$. Fix such a u. Given any $z \in \mathbb{F}_q^{k-1}$, define $X' = (X | F(X, u) = z)$. $\mathrm{Supp}(X')$ is not empty by our choice of u. Moreover, since the conditioning $F(X, u) = z$ simply adds $k - 1$ affine constraints on $\mathrm{Supp}(X)$, $\mathrm{Supp}(X')$ is an affine subspace of dimension at least 1. Thus, X' is a subsource of X that is also an affine source with min-entropy at least $\log q$. Since $\Pr[F(X', u) = z) = 1]$, this proves the claim. □

Theorem 5.36. *Fix any sufficiently large prime power q and any integers n, k such that $q \geq n^{18}$ and $2 \leq k < n$. There is a poly$(n, \log q)$-time computable zero-error disperser for $(n, k)_q$-affine sources $D : \mathbb{F}_q^n \mapsto \{0,1\}^{(k-1) \cdot \log q}$.*[5]

Proof. Use Theorem 5.15 with D from Corollary 5.5 as D', and F from Corollary 5.6. □

5.7 Open Problems

2-sources. One of the most important open problems in this area is giving constructions of extractors for entropy threshold $k = o(n)$. Such constructions are not known even for $m = 1$ and large error ϵ.

There are explicit constructions of zero-error dispersers with $k = n^{o(1)}$ [5]. However, these dispersers only output one bit. A consequence of Corollary 5.2 is that improving the output length in these constructions to $\Theta(\log n)$ bits will allow our composition techniques to achieve output length $m = \Omega(k)$.

Another intriguing problem is that for the case of zero-error (or strongly hitting) dispersers we do not know whether the existential results proven via the probabilistic method achieve the best possible parameters. More precisely, a straightforward application of the probabilistic method gives zero-error 2-source dispersers which on entropy threshold $2 \cdot k$ output $m = k - \log(n-k) - O(1)$ bits. On the other hand the lower bounds of [48, 49] can be used to show that any zero-error 2-source disperser with entropy threshold $2 \cdot k$ has $m \leq k + O(1)$.[6]

---

[5]When we say that D is poly$(n, \log q)$-time computable we mean that computing D requires poly(n) field operations in $\mathbb{F}_q$. Thus, assuming we have a representation of $\mathbb{F}_q$ in which addition and multiplication can be done in poly$(\log q)$ time (which is true for all standard representations), we get that D is poly$(n, \log q)$-time computable.

[6]Radhakrishnan and Ta-Shma [49] show that any seeded disperser $D : \{0,1\}^n \times \{0,1\}^t \to \{0,1\}^m$ that is nontrivial in the sense that $m \geq t+1$ has $t \geq \log(1/\epsilon) - O(1)$. A zero-error 2-source disperser D' with entropy threshold k can be easily transformed into a seeded disperser with seed length $t = k$ by setting $D(x, y) = D'(x, y')$ where y' is obtained by padding the k-bit-long "seed" y with $n - k$ zeroes. The bound follows as D' has error smaller than 2^{-m}.

O(1)-sources, rainbows and implicit probe search. When allowing ℓ-sources for $\ell = O(1)$ we give constructions of zero-error dispersers which on entropy threshold $k = n^{\Omega(1)}$ achieve output length $m = \Omega(k)$. An interesting open problem is to try to improve the entropy threshold. As explained in Subsection 5.4.5, this immediately implies improved implicit probe search schemes.

Bit-fixing sources. We give constructions of zero-error dispersers which on entropy threshold k achieve $m = \Omega(k)$. A straightforward application of the probabilistic method gives zero-error dispersers with $m = k - \log n - o(\log n)$. We do not know how to match these parameters with explicit constructions.

Affine sources. We constructed a subsource hitter for affine sources over relatively large fields (that is, $q = n^{\Theta(1)}$). It is interesting to try and construct subsource hitters for smaller fields.

Finally, it is also natural to ask whether our composition approach applies to other classes of sources.

Appendix A

Sampling and Partitioning

In this appendix we give constructions of samplers and prove Lemmas 2.16, 2.17 and 2.18.

A.1 Sampling Using ℓ-wise Independence

Bellare and Rompel [6] gave a sampler construction based on ℓ-wise independent variables. We use a twist on their method: Suppose we are aiming to hit k/r bits when given a subset S of size k. We generate ℓ-wise independent variables $Z_1, \dots, Z_n \in [r]$ and define $T = \{i | Z_i = 1\}$. It follows that with high probability $S \cap T$ is of size approximately k/r. This is stated formally in the following lemma. (We explain the difference between this method and that of [6] in Remark A.2.)

Lemma A.1. *For every integers n, k, r, t such that $r \le k \le n$ and $6 \log n \le t \le \frac{k \log n}{20r}$ there is an explicit $(n, k, \frac{1}{2} \cdot \frac{k}{r}, 3 \cdot \frac{k}{r}, 2^{-\Omega(t/\log n)})$-sampler which uses a seed of t random bits.*

Before proving this lemma we show that Lemma 2.16 is a special case.

Proof. (of Lemma 2.16) We use Lemma A.1 with the parameters n, k and $r = \frac{3k}{n^{1/2+\gamma}}$, $t = \alpha \cdot n^{2\gamma}$. We need to check that $6 \log n \le t \le \frac{k \log n}{20r}$. Clearly, $t \ge 6 \log n$ (for a large enough n depending on α and γ). On the other hand,

$$\frac{k \log n}{20r} = \frac{n^{1/2+\gamma} \log n}{60} \ge \alpha \cdot n^{2\gamma} = t$$

(for a large enough n depending on α and γ). Thus, applying Lemma A.1, we get an $(n, k, k/2r, 3k/r, \delta)$-sampler $Samp : \{0,1\}^t \to P([n])$ where

$$\delta = 2^{-\Omega(t/\log n)} = 2^{-\Omega(\alpha \cdot n^{2\gamma}/\log n)} = 2^{-\Omega(\alpha \cdot n^{\gamma})}$$

(for a large enough n depending on α and γ). $\square$

A. Gabizon, *Deterministic Extraction from Weak Random Sources*, Monographs in Theoretical Computer Science. An EATCS Series, DOI 10.1007/978-3-642-14903-0,

We need the following tail inequality for ℓ-wise independent variables due to Bellare and Rompel [6].

Theorem A.1 ([6]). *Let $\ell \geq 6$ be an even integer. Suppose that $X_1, \ldots, X_n$ are ℓ-wise independent random variables taking values in $[0,1]$. Let $X = \sum_{1 \leq i \leq n} X_i$ and $\mu = E(X)$, and let $A > 0$. Then*

$$\Pr[|X - \mu| \geq A] \leq 8 \left(\frac{\ell\mu + \ell^2}{A^2} \right)^{\ell/2}.$$

We now prove Lemma A.1.

Proof. (of Lemma A.1) Let ℓ be the largest even integer such that $\ell \log n \leq t$ and let $q = \lfloor \log r \rfloor \leq \log n$. There are constructions which use $\ell \log n \leq t$ random bits to generate n random variables $Z_1 \ldots, Z_n \in \{0,1\}^q$ that are ℓ-wise independent [13]. The sampler generates such random variables. Let $a \in \{0,1\}^q$ be some fixed value. We define a random variable $T = \{i | Z_i = a\}$. Let $S \subseteq [n]$ be some subset of size k. For $1 \leq i \leq n$ we define a boolean random variable X_i such that $X_i = 1$ if $Z_i = a$. Let $X = |S \cap T| = \sum_{i \in S} X_i$. Note that $\mu = E(X) = k/2^q$ and that the random variables $X_1, \ldots, X_n$ are ℓ-wise independent. Applying Theorem A.1 with $A = k/2r$ we get that

$$\Pr[|X - \mu| \geq A] \leq 8 \left(\frac{\ell k/2^q + \ell^2}{A^2} \right)^{\ell/2}.$$

Note that

$$\{|X - \mu| < A\} \subseteq \left\{ \frac{k}{2^q} - A < X < \frac{k}{2^q} + A \right\} \subseteq \left\{ \frac{k}{r} - A < X < \frac{2k}{r} + A \right\}$$
$$\subseteq \{k_{min} \leq X \leq k_{max}\}$$

for $k_{min} = k/2r$ and $k_{max} = 3k/r$. Note that $\ell \leq \frac{t}{\log n} \leq \frac{k}{20r}$. We conclude that

$$\Pr[k_{min} \leq |S \cap T| \leq k_{max}] \geq 1 - 8 \left(\frac{\ell \frac{k}{2^q} + \ell^2}{(\frac{k}{2r})^2} \right)^{\ell/2} \geq 1 - 8 \left(\frac{4r^2 (\frac{2\ell k}{r} + \frac{\ell k}{20r})}{k^2} \right)^{\ell/2}$$
$$\geq 1 - 8 \left(\frac{10\ell r}{k} \right)^{\ell/2} \geq 1 - 2^{-(\ell/2+3)} \geq 1 - 2^{-\Omega(t/\log n)}.$$

□

Remark A.2. *We remark that this construction is different from the common way of using ℓ-wise independence for sampling [6]. The more common way is to take n/r random variables $V_1, \ldots, V_{n/r} \in [n]$ which are ℓ-wise independent and sample the multi-set $T = \{V_1, \ldots, V_{n/r}\}$. The expected size of the multi-set $|S \cap T|$ is k/r and one gets the same probability of success*

$\delta = 2^{-\Omega(\ell)}$ by the tail inequality of [6]. The two methods require roughly the same number of random bits. Nevertheless, the method of Lemma A.1 has the following advantages:

- *It can also be used for partitioning.*
- *The method used in Lemma A.1 guarantees that T is a set whereas the standard method may produce a multi-set.*
- *The method used in Lemma A.1 can be derandomized and use much fewer bits (at least for small r and large δ). More precisely, suppose that $r \leq \log n$ and say $\ell = 2$. In this range of parameters, one can use $O(\log\log n)$ random bits to generate n variables $Z_1, \ldots, Z_n \in \{0,1\}^{\log r}$ which are $(1/\log n)$-close to being pairwise independent. Thus, the same technique can be used to construct more randomness efficient samplers (at the cost of having a larger error parameter δ.) We use this idea in Section A.2. We remark that in the case of the standard method no savings can be made as it requires variables Z_i over $\{0,1\}^{\log n}$ and even sampling one such variable requires $\log n$ random bits.*

A.2 Sampling and Partitioning Using Fewer Bits

We now derandomize the construction of Lemma A.1 to give schemes which use only $O(\log k)$ bits and prove Lemmas 2.17 and 2.18. These two lemmas follow from the following more general lemma.

Lemma A.3. *Fix any integer $n \geq 16$. Let k be an integer such that $k \leq n$. Let r satisfy $r \leq k$. Let r' be the power of 2 that satisfies $(1/2)r < r' \leq r$. Let $\epsilon > 0$ satisfy $1/kr \leq \epsilon \leq 1/8r$. We can use $7\log r + 3(\log\log n + \log(1/\epsilon))$ random bits to explicitly partition $[n]$ into r' sets $T_1, \ldots, T_{r'}$ such that for any $S \subseteq [n]$ where $|S| = k$*

$$\Pr(\forall i, \quad k/2r \leq |T_i \cap S| \leq 3k/r) \geq 1 - O(\epsilon \cdot r^3).$$

We prove Lemma A.3 in the next section. We now explain how the two lemmas follow from Lemma A.3.

Proof. (of Lemma 2.18) Set $b = \alpha/38$. Use Lemma A.3 with the parameters $r = k^b$ and $\epsilon = k^{-4b}$ to obtain a partition $T_1, \ldots, T_{r'}$ of $[n]$ where $(1/2)r < r' \leq r$ is a power of 2.

To use Lemma A.3 with these parameters we need $7\log r + 3(\log\log n + \log(1/\epsilon)) = 7\log k^b + 3(\log\log n + \log k^{4b})$ random bits. We want to use at most $\alpha \cdot \log k$ bits.
Set $c = 6/\alpha$. Since we assume that $k \geq \log^c n$,

$$(\alpha/2)\log k \geq (\alpha/2)(6/\alpha)\log\log n = 3\log\log n.$$

So now we need

$$(\alpha/2)\log k \geq 7\log k^b + 3\log k^{4b} = b(7+12)\log k.$$

Or, equivalently,

$$b \leq \alpha/38.$$

Set $e = 1 - b$. So $k/2r = k^e/2$ and $3k/r = 3 \cdot k^e$. Note that $e > 1/2$ as required.
Using Lemma A.3,

$$\Pr(\forall i, \quad k^e/2 \leq |T_i \cap S| \leq 3 \cdot k^e) \geq 1 - O(\epsilon \cdot r^3) = 1 - O(k^{-b}).$$

□

Lemma 2.17 easily follows from Lemma 2.18.

Proof. (of Lemma 2.17) Use Lemma 2.18 with the parameters n, k and α to obtain a partition of $[n]$ $T_1, \ldots, T_m$ and take T_1 as the sample. It is immediate that the required parameters are achieved. □

Proof of Lemma A.3

The sampler construction in Lemma A.1 relied on random variables $Z_1, \ldots, Z_n \in [r]$, which are ℓ-wise independent. We now show that we can derandomize this construction and get a (weaker) sampler by using $Z_1, \ldots, Z_n$ which are only *pairwise ϵ-dependent.* Naor and Naor [44] (and later Alon et al.[2]) gave constructions of such variables using very few random bits. This allows us to reduce the number of random bits required for sampling and partitioning.

The following definition formalizes a notion of limited independence, slightly more general than the one discussed above:

Definition A.4 (ℓ-wise ϵ-dependent variables)**.** *Let D be a distribution. We say that the random variables $Z_1, \ldots, Z_n$ are ℓ-wise ϵ-dependent according to D if for every $M \subseteq [n]$ such that $|M| \leq \ell$, the distribution Z_M (that is, the joint distribution of the Z_is such that $i \in M$) is ϵ-close to the distribution $D^{\otimes |M|}$, i.e., the distribution of $|M|$ independent random variables chosen according to D. We sometimes omit D when it is the uniform distribution. Random bit variables $B_1, \ldots, B_n$ are ℓ-wise ϵ-dependent with mean p if they are ℓ-wise ϵ-dependent according to the distribution $D = (1-p, p)$ on $\{0,1\}$.*

We need two properties about ℓ-wise ϵ-dependent variables: That they can be generated using very few random bits and that their sum is concentrated around the expectation. The first property is proven in Lemma A.5 and the second in Lemma A.6.

The following theorem states that ℓ-wise ϵ-dependent bit variables can be generated by very few random bits.

Theorem A.2 ([2]). [1]*For any* $n \geq 16$, $\ell \geq 1$ *and* $0 < \epsilon < 1/2$, ℓ*-wise* ϵ*-dependent bits* $B_1, \ldots, B_n$ *can be generated using* $3(\ell + \log\log n + \log(1/\epsilon))$ *truly random bits.*

We can generate pairwise ϵ-dependent variables in larger domains using ℓ-wise ϵ-dependent bit variables.[2]

Lemma A.5. *Let* $r < n$ *be a power of* 2. *For any* $n \geq 16$ *and* $0 < \epsilon < 1/2$, *we can generate pairwise* ϵ*-dependent variables* $Z_1, \ldots, Z_n \in [r]$ *using* $7 \log r + 3(\log\log n + \log(1/\epsilon))$ *truly random bits.*

Proof. Using Theorem A.2, we generate $2\log r$-wise ϵ-dependent bit variables $B_1, \ldots, B_{n \log r}$ using $3(2\log r + \log\log(n \log r) + \log(1/\epsilon)) \leq 7\log r + 3(\log\log n + \log(1/\epsilon))$ bits. We partition the B_is into n blocks of size $\log r$ and interpret the ith block as a value $Z_i \in [r]$.

The joint distribution of the bits in any block or 2 blocks is ϵ-close to uniform. Therefore, the Z_is are pairwise ϵ-dependent. □

In the following lemma, we use Chebychev's inequality to show that the sum of pairwise ϵ-dependent bit variables is close to its expectation with high probability.

Lemma A.6. *Let* p *satisfy* $0 < p < 1$. *Let* $\epsilon > 0$ *satisfy* $p/k \leq \epsilon \leq p/4$. *Let* $B_1, \ldots, B_k$ *be pairwise* ϵ*-dependent bit variables with mean* p. *Let* $B = \sum_{i=1}^{k} B_i$.
Then

$$\Pr(|B - pk| > pk/2) = O(\epsilon/p^2).$$

Proof. Using linearity of expectation we get $|E(B) - pk| \leq \epsilon k$.
Therefore,

$$\Pr(|B - pk| > pk/2) \leq \Pr(|B - E(B)| > pk/2 - \epsilon k).$$

So it's enough to bound

$$\Pr(|B - E(B)| > pk/2 - \epsilon k).$$

Fix any $i, j \in [k]$ where $i \neq j$. The covariance of B_i and B_j will be small since they are almost independent:

$$cov(B_i, B_j) = E(B_i \cdot B_j) - E(B_i)E(B_j)$$

$$= \Pr(B_i = 1; B_j = 1) - \Pr(B_i = 1)\Pr(B_j = 1)$$

[1]The theorem is stated a bit differently and only for odd ℓ in ([2]), but this form is easily deduced from Theorem 3 in that paper by observing that $(\ell + 1)$-wise ϵ-dependence implies ℓ-wise ϵ-dependence.

[2]Actually, a construction of such (and more general types of) variables already appears in [23].

$$\leq (p^2 + \epsilon) - (p - \epsilon)^2 = (1 + 2p - \epsilon)\epsilon \leq 3\epsilon$$

(where the second equality is because B_i and B_j are bit variables)

Therefore, the variance of B won't be too large:

$$Var(B) = \sum_i Var(B_i) + \sum_{i \neq j} cov(B_i, B_j) \leq (p + \epsilon)k + 3\epsilon k^2 \leq pk + 4\epsilon k^2.$$

Therefore, by Chebychev's inequality,

$$\Pr(|B - E(B)| > pk/2 - \epsilon k) < \frac{pk + 4\epsilon k^2}{(pk/2 - \epsilon k)^2}.$$

We required that $\epsilon \leq p/4$, and therefore

$$\leq \frac{pk + 4\epsilon k^2}{(pk/4)^2} = O(1/pk) + O(\epsilon/p^2) = O(\epsilon/p^2)$$

(where the last equality follows by the requirement that $\epsilon \geq p/k$). □

Now we can easily prove Lemma A.3.

Proof. (of Lemma A.3) Let r' be the power of 2 in the statement of the lemma. Using Lemma A.5, we generate pairwise ϵ-dependent $Z_1, \ldots, Z_n \in [r']$. For $1 \leq i \leq r'$, we define $T_i = \{j | Z_j = i\}$.

Assume, w.l.o.g., that $S = \{1, \ldots, k\}$. Given $i \in [r']$, define the bit variables $B_1, \ldots, B_k$ by $B_j = 1 \Leftrightarrow Z_j = i$. It is easy to see that the B_js are pairwise 2ϵ-dependent with mean $1/r'$. Define $C_i = \sum_{j=1}^{k} B_j$.

Note that $C_i = |T_i \cap S|$. Notice that $1/r'$ and 2ϵ satisfy the requirements in Lemma A.6.

Using Lemma A.6,

$$\Pr(|C_i - k/r'| > k/2r') = O(\epsilon \cdot (r')^2) = O(\epsilon \cdot r^2).$$

Using the union bound,

$$\Pr(\exists i \ s.t \ |C_i - k/r'| > k/2r') = O(\epsilon \cdot r^3).$$

Thus, we can obtain a partition $T_1, \ldots, T_{r'}$ of $[n]$ such that, with probability at least $1 - O(\epsilon \cdot r^3)$,

$$\forall i \ k/2r' \leq |T_i \cap S| \leq 3k/2r',$$

which implies that with at least the same probability,

$$\forall i \ k/2r \leq |T_i \cap S| \leq 3k/r.$$

□

Appendix B

Basic Notions from Algebraic Geometry

In Section 4.5 we use a theorem of Bombieri [8] regarding character sums over curves. The very statement, let alone the applicability of Bombieri's theorem, requires some basic notions from algebraic geometry. In this appendix, we give some basic background necessary for stating the theorem and applying it as done in Section 4.5. The main issue in Section 4.5 is to show that the varieties that come up there are suitable for the theorem. Specifically, we need to show that these varieties are indeed curves, i.e., have dimension 1, and that their 'degree' is not too large. (All these terms will be defined formally). For this purpose, we need some lemmas regarding the dimension and degree of intersections of varieties. Another issue is that Bombieri's theorem is stated for projective curves while we want to apply it on affine curves. For this purpose, we need some lemmas on the relations between affine and projective varieties. We note that all these issues are standard. We stress that this section is far from a full introduction to basic algebraic geometry. For a very accessible introduction we recommend [17], of which most the definitions and notations in this section follow.
Throughout this section $\mathbb{F}$ will always denote an algebraically closed field.

B.1 Affine and Projective Varieties

The basic objects of study in algebraic geometry are the sets of solutions to a system of polynomial equations. Such a set is called a *variety.* We now formally define affine space and affine varieties.

Definition B.1 (affine space). *We define n-dimensional affine space over* $\mathbb{F}$ *as*[1]

$$\mathbb{F}^n \triangleq \{(a_1, \ldots, a_n) \mid a_i \in \mathbb{F}\}.$$

[1]In most textbooks in algebraic geometry the notation $\mathbb{A}^n$ is used rather than $\mathbb{F}^n$. However, in [17], which we are following, $\mathbb{F}^n$ is used.

Definition B.2 (affine variety). *Let $f_1, \ldots, f_s$ be polynomials in $\mathbb{F}[x_1, \ldots, x_n]$. We set*

$$\mathbf{V}(f_1, \ldots, f_s) = \{(a_1, \ldots, a_n) \in \mathbb{F}^n \mid \forall\, 1 \leq i \leq s\ f_i(a_1, \ldots, a_n) = 0\}.$$

We call $\mathbf{V}(f_1, \ldots, f_s)$ the affine variety *defined by $f_1, \ldots, f_s$. A subset $V \subseteq \mathbb{F}^n$ is an* affine variety *if $V = \mathbf{V}(f_1, \ldots, f_s)$ for some set of polynomials $f_1, \ldots, f_s \in \mathbb{F}[x_1, \ldots, x_n]$. We say that V is* reducible *if it can be written as $V = V_1 \cup V_2$ where the V_is are affine varieties such that $V \neq V_1, V_2$. Otherwise, we say that V is* irreducible.[2]

As a simple example of an affine variety, take $V = \mathbf{V}(x_1 \cdot x_2) \subseteq \mathbb{F}^2$. Note that V is reducible as it is the union of the varieties $V_1 = \mathbf{V}(x_1)$ and $V_2 = \mathbf{V}(x_2)$, i.e., the sets $\{(0, x_2)|x_2 \in \mathbb{F}\}, \{(x_1, 0)|x_1 \in \mathbb{F}\} \subseteq \mathbb{F}^2$. It can be shown that V_1 and V_2 are irreducible. Note that this is not a disjoint union as $V_1 \cap V_2 = (0, 0)$.

Though affine space and affine varieties seem to be the natural objects we want to investigate, it turns out to be very useful to work in *projective space*. Intuitively, projective space is affine space extended with additional 'extra points'. This intuition may not be clear from the following definition but will become clearer later on.

Definition B.3 (projective space). *We define an equivalence relation $\sim$ over $\mathbb{F}^{n+1} \backslash \{0\}$ by setting*

$$(x_0, \ldots, x_n) \sim (y_0, \ldots, y_n)$$

if and only if there exists a nonzero $\lambda \in \mathbb{F}$ such that $(x_0, \ldots, x_n) = (\lambda \cdot y_0, \ldots, \lambda \cdot y_n)$. We define n-dimensional projective space $\mathbb{P}^n$ over $\mathbb{F}$ to be the set of all equivalence classes of $\sim$. Thus,

$$\mathbb{P}^n = (\mathbb{F}^{n+1} - \{0\}) / \sim .$$

Each non-zero $(n+1)$-tuple $(x_0, \ldots, x_n) \in \mathbb{F}^n$ defines a point $p \in \mathbb{P}^n$. We say that $(x_0, \ldots, x_n)$ are homogenous coordinates *of p.*

We say that a polynomial $f \in \mathbb{F}[x_0, \ldots, x_n]$ is *homogenous* if all of its monomials have the same total degree. It is easy to see that for a homogenous polynomial f of total degree d and any nonzero $\lambda \in \mathbb{F}$

$$f(\lambda \cdot a_0, \ldots, \lambda \cdot a_n) = \lambda^d f(a_0, \ldots, a_n).$$

In particular, $f(\lambda \cdot a_0, \ldots, \lambda \cdot a_n) = 0$ if and only if $f(a_0, \ldots, a_n) = 0$. Thus, the set of 'zeros' of f is a well-defined object in $\mathbb{P}^n$.

This leads to the following definition.

---

[2] In many textbooks, the term variety always means an irreducible variety and general varieties are called *algebraic sets*.

Definition B.4 (projective variety). *Let $f_1, \ldots, f_s \in \mathbb{F}[x_0, \ldots, x_n]$ be homogenous polynomials. We set*

$$\mathbf{V}(f_1, \ldots, f_s) = \{(a_0, \ldots, a_n) \in \mathbb{P}^n \mid \forall\, 1 \le i \le s\ f_i(a_0, \ldots, a_n) = 0\}.$$

A subset $V \subseteq \mathbb{P}^n$ is a projective variety *if $V = \mathbf{V}(f_1, \ldots, f_s)$ for some set of homogenous polynomials $f_1, \ldots, f_s \in \mathbb{F}[x_0, \ldots, x_n]$. We say that V is* reducible *if it can be written as $V = V_1 \cup V_2$, where the V_is are projective varieties such that $V \neq V_1, V_2$. Otherwise, we say that V is* irreducible.

An important basic property of (affine and projective) varieties is that they decompose into irreducible varieties in a unique way. Thus, we can speak unambiguously about the irreducible components of a variety.

Proposition B.1. *-[[17], Chapter 4, §6, Theorem 4, and Chapter 8, §3, Theorem 6] We say that $V = V_1 \cup \ldots \cup V_m$ is a* minimal decomposition *of V if $V_i \not\subseteq V_j$ for every $i \neq j$. Let V be an affine (projective) variety. Then V has a minimal decomposition*

$$V = V_1 \cup \ldots \cup V_m$$

where the V_is are irreducible affine (projective) varieties. Furthermore, this minimal decomposition is unique up to the order in which $V_1, \ldots, V_m$ are written.

B.2 Varieties and Ideals

An affine variety is essentially a geometric object — a set of points in the space $\mathbb{F}^n$. A fundamental idea in algebraic geometry is to relate a variety to an algebraic object. This algebraic object will be the set of all polynomials that vanish on the variety. It is easy to see that this set of polynomials forms an ideal in the ring $\mathbb{F}[x_1, \ldots, x_n]$. First we recall some basic facts and notation regarding ideals in $\mathbb{F}[x_1, \ldots, x_n]$. For $f_1, \ldots, f_s \in \mathbb{F}[x_1, \ldots, x_n]$ we denote by $< f_1, \ldots, f_s >$ the ideal generated by $f_1, \ldots, f_s$. That is,

$$< f_1, \ldots, f_s > \triangleq \left\{ \sum_{i=1}^{s} g_i \cdot f_i \mid \forall\, 1 \le i \le s\ g_i \in \mathbb{F}[x_1, \ldots, x_n] \right\}.$$

By the Hilbert Basis Theorem (see [17], Chapter 2, §5) every ideal $I \subset \mathbb{F}[x_1, \ldots, x_n]$ is *finitely generated*, i.e., $I = < f_1, \ldots, f_s >$ for some $f_1, \ldots, f_s \in \mathbb{F}[x_1, \ldots, x_n]$. For an ideal $I = < f_1, \ldots, f_s >$, it is easy to see that a point $(a_1, \ldots, a_n) \in \mathbb{F}^n$ is a zero of every $f \in I$ if and only if it is a zero of $f_1, \ldots, f_s$.

Definition B.5 (affine varieties and ideals). *For an affine variety $V \subseteq \mathbb{F}^n$ we define $\mathbf{I}(V)$ to be the ideal of all polynomials f such that $f(a_1, \ldots, a_n) = 0$ for every $(a_1, \ldots, a_n) \in V$. For an ideal $I = < f_1, \ldots, f_s > \subseteq \mathbb{F}[x_1, \ldots, x_n]$ we define $\mathbf{V}(I) \subseteq \mathbb{F}^n$ to be the affine variety $\mathbf{V}(I) = \{(a_1, \ldots, a_n) \mid f(a_1, \ldots, a_n) = 0,\ \forall f \in I\} = \mathbf{V}(f_1, \ldots, f_s)$.*

Before making the corresponding definitions for projective varieties we will need some terminology. We remarked above that it makes sense to ask whether a homogenous polynomial $f \in \mathbb{F}[x_0, \ldots, x_n]$ vanishes at a point $p \in \mathbb{P}^n$. For a non-homogenous polynomial $f \in \mathbb{F}[x_0, \ldots, x_n]$ we say that $f(p) = 0$ for $p \in \mathbb{P}^n$ if $f(a_0, \ldots, a_n) = 0$ for all representatives $(a_0, \ldots, a_n)$ of p.

We say that an ideal $I \subseteq \mathbb{F}[x_0, \ldots, x_n]$ is *homogenous* if it is generated by a set of homogenous polynomials, i.e., $I = < f_1, \ldots, f_s >$ where $f_1, \ldots, f_s$ are homogenous. We can now make the following definitions.

Definition B.6 (projective varieties and homogenous ideals). *For a projective variety $X \subseteq \mathbb{P}^n$ we define $\mathbf{I}(X)$ to be the ideal of all polynomials f with $f(p) = 0$ for every $p \in X$. It can be shown that $\mathbf{I}(X)$ is always a homogenous ideal. For a homogenous ideal $I \subseteq \mathbb{F}[x_0, \ldots, x_n]$ we define $\mathbf{V}(I) \subseteq \mathbb{P}^n$ to be the projective variety of all points $p \in \mathbb{P}^n$ that are zeros of all polynomials $f \in I$. If $I = < f_1, \ldots, f_s >$ for homogenous polynomials $f_1, \ldots, f_s$ then it can be shown that $\mathbf{V}(I) = \mathbf{V}(f_1, \ldots, f_s)$.*

One reason the correspondence between ideals and varieties is useful is that operations on ideals have simple corollaries in terms of the corresponding varieties. We need the following fact about intersections of ideals.

Proposition B.2 ([17], Chapter 4, §3, Theorem 15, and Chapter 8, §3, Exercise 7). *Let I_1, I_2 be ideals in $\mathbb{F}[x_1, \ldots, x_n]$ or homogenous ideals in $\mathbb{F}[x_0, \ldots, x_n]$. Then*

$$\mathbf{V}(I_1 \cap I_2) = \mathbf{V}(I_1) \cup \mathbf{V}(I_2).$$

B.3 The Dimension and Degree of a Variety

There are several equivalent definitions of the dimension and degree of a variety (degree is defined only for projective varieties). Here we define dimension and degree in terms of the Hilbert polynomial of a variety. First we need to define the Hilbert function and Hilbert polynomial of an ideal. The definitions are taken from [17].

We say that an ideal I is a *monomial ideal* if it is generated by a set of monomials.[3]. For example, $I = < x_1, x_2^2 >$ is a monomial ideal. We first define the Hilbert function for monomial ideals.

Definition B.7 (Hilbert function of a monomial ideal). *Let I be a monomial ideal in $\mathbb{F}[x_1, \ldots, x_n]$. The* affine Hilbert function *of I, denoted by ${}^aHF_I(s)$, is a function on non-negative integers defined by ${}^aHF_I(s)$ = number of monic monomials in $\mathbb{F}[x_1, \ldots, x_n]$ of degree at most s not contained in I. Similarly, let I be a homogenous monomial ideal in $\mathbb{F}[x_0, \ldots, x_n]$. The* Hilbert function

[3]By Dickson's Lemma ([17], Chapter 2, §4, Theorem 5), if I is a monomial ideal it can always be generated by a finite set of monomials.

of I, denoted by $HF_I(s)$, is a function on non-negative integers defined by $HF_I(s)$ = number of monic monomials in $\mathbb{F}[x_0,\ldots,x_n]$ of degree exactly s *not contained in I.*

Roughly speaking, for a monomial ideal I the monomials not in I are a basis for the space of polynomials that are 'different modulo I'. Thus, ${}^aHF_I(s)$ is the dimension of the space of such polynomials of degree at most s. This is the idea behind the definition of the Hilbert function for general ideals. First we need some notation. For a subset of polynomials $V \subseteq \mathbb{F}[x_1,\ldots,x_n]$ and a non-negative integer s, we denote by $V_{\leq s} \subseteq \mathbb{F}[x_1,\ldots,x_n]$ the set of polynomials in V of (total) degree at most s. For example, $\mathbb{F}[x_1,\ldots,x_n]_{\leq s}$ is the set of all polynomials of degree at most s. Similarly, for a subset $V \subseteq \mathbb{F}[x_0,\ldots,x_n]$ we denote by $V_s \subseteq \mathbb{F}[x_0,\ldots,x_n]$ the set of all polynomials in V of degree *exactly* s. Note that if $V \subseteq \mathbb{F}[x_1,\ldots,x_n]$ is a linear subspace, then so are $V_{\leq s}$ and V_s. In particular if $I \subseteq \mathbb{F}[x_1,\ldots,x_n]$ is an ideal, then it is also a linear subspace, and so is $V_{\leq s}$. We recall a basic notion for linear algebra: For subspaces $W \subseteq V \subseteq \mathbb{F}[x_1,\ldots,x_n]$ we denote by V/W the *quotient space* of equivalence classes of V over W. That is, we define an equivalence relation $\sim$ over V by $v \sim v' \leftrightarrow v - v' \in W$ and let V/W be the space of these equivalence classes. We can now make the following definition.

Definition B.8 (Hilbert function of a general ideal). *Let I be an ideal in $\mathbb{F}[x_1,\ldots,x_n]$. The* affine Hilbert function *of I, denoted by ${}^aHF_I(s)$, is defined as*
${}^aHF_I(s) \triangleq dim\,(\mathbb{F}[x_1,\ldots,x_n]_{\leq s}/I_{\leq s})$.
Let I be a homogenous ideal in $\mathbb{F}[x_0,\ldots,x_n]$; the Hilbert function *of I, denoted $HF_I(s)$, is defined as $HF_I(s) \triangleq dim\,(\mathbb{F}[x_0,\ldots,x_n]_s/I_s)$.*

It can be shown that for large enough input s, the Hilbert Function coincides with a polynomial.

Theorem B.1 (see [17] Chapter 9, §3).

1. *Let I be an ideal in $\mathbb{F}[x_1,\ldots,x_n]$. There exists a polynomial ${}^aHP_I(s)$ such that for large enough s, ${}^aHP_I(s) = {}^aHF_I(s)$. We call ${}^aHP_I(s)$ the* affine Hilbert polynomial *of I.*

2. *Let I be a homogenous ideal in $\mathbb{F}[x_0,\ldots,x_n]$. There exists a polynomial $HP_I(s)$ such that for large enough s, $HP_I(s) = HF_I(s)$. We call $HP_I(s)$ the* Hilbert polynomial *of I.*

Let $V \subseteq \mathbb{F}^n$ be an affine variety with $I = \mathbf{I}(V)$. Let's try to see why it could make sense to define the dimension of a variety in terms of the affine Hilbert polynomial of I. Since I is exactly the set of polynomials that vanish on V, polynomials $f, g \in \mathbb{F}[x_1,\ldots,x_n]$ are identical on V if and only if $f - g \in I$. It follows that $\mathbb{F}[x_1,\ldots,x_n]/I$ is exactly the space of polynomial functions from V to $\mathbb{F}$. Now recall that for a linear subspace $A \subseteq \mathbb{F}^n$, the dimension of A can be defined as the dimension of the space of linear functions from A to $\mathbb{F}$. Similarly, we could try to define the dimension of V as the dimension of the space of *polynomial* functions from V to $\mathbb{F}$, i.e., the dimension of $\mathbb{F}[x_1,\ldots,x_n]/I$. However, since the polynomials in this space have unbounded degree, $\mathbb{F}[x_1,\ldots,x_n]/I$ has infinite dimension. Instead, we can take an 'asymptotic' approach and define the dimension of V by how fast this space grows as we increase the degree of the polynomials. More accurately, we can define $dim(V)$ by how fast ${}^aHP_I(s) = dim(\mathbb{F}[x_1,\ldots,x_n]_{\leq s}/I_{\leq s})$ grows as s increases. This corresponds to the degree of ${}^aHP_I(s)$.

Definition B.9 (dimension of a variety). *Let $V \subseteq \mathbb{F}^n$ be an affine variety and let $I = \mathbf{I}(V)$. The* dimension *of V, denoted $dim(V)$, is defined to be the degree of ${}^aHP_I(s)$. Let $V \subseteq \mathbb{P}^n$ be a projective variety and let $I = \mathbf{I}(V)$. The* dimension *of V is defined to be the degree of $HP_I(s)$.*

To gain intuition on the above definition, it is helpful to see how it coincides with the definition of dimension for a linear subspace. Take for example the subspace $V \subseteq \mathbb{F}^n$ defined by the constraints $\{x_1 = 0,\ x_2 = 0\}$. Then $I \triangleq \mathbf{I}(V) =< x_1, x_2 >$ and the monomials *not* in I are exactly the monomials $x_3^{a_3} \cdots x_n^{a_n}$, where $a_3,\ldots,a_n$ are non-negative integers. In particular, the number of such monomials of degree at most s is $\binom{n-2+s}{n-2}$, which is a degree $n-2$ polynomial in s. Therefore, since I is a monomial ideal by the definition above, $dim(V) = deg(HP_I(s)) = n-2$.

The following property of the dimension of a variety will be very useful for us later on.

Proposition B.3 ([17], Chapter 9, §4 Corollary 9). *Let V be an affine or projective variety. The dimension of V is equal to the maximum of the dimensions of its irreducible components.*

We now define the *degree* of a projective variety (degree is not defined for affine varieties).

Definition B.10 (degree of a variety). *The* degree *of V, denoted by $deg(V)$, is defined to be the leading coefficient of $HP_I(s)$ multiplied by $dim(V)!$.*

Though not immediate from the definition, it can be shown that the degree is always a non-negative integer. To gain intuition on the above definition, let us see how it coincides with the definition of degree for a univariate polynomial. For simplicity of the example we'll assume degree is defined for an affine variety V in a similar way to projective varieties. That is, $deg(V)$ is the leading coefficient of the affine Hilbert polynomial of $\mathbf{I}(V)$ times $dim(V)!$. Let $I \subseteq \mathbb{F}[x_1]$ be the ideal $< x_1^3 - 1 >$. It can be shown that $I = \mathbf{I}(V)$ where $V = \mathbf{V}(x_1^3 - 1) \in \mathbb{F}$, i.e., V is simply the roots of $x_1^3 - 1$ and $|V| = 3$ (since $\mathbb{F}$ is algebraically closed). Furthermore, it can be seen that $\{1, x_1, x_1^2\}$ is a basis for $k[x_1]/I$. Hence, $HP_I(s)$ is simply the constant 3, and therefore $dim(V) = deg(HP_I(s)) = 0$ and $deg(V) = 3 \cdot 0! = 3$. Thus $deg(V)$ bounds the size of V. It can be shown that $deg(V)$ always bounds $|V|$ when V is a projective variety of finite size.

B.4 The Projective Closure of an Affine Variety

We call an affine (projective) variety of dimension 1 an affine (projective) curve. As mentioned above, in Section 4.5 we use a theorem of Bombieri[8] for affine curves while in [8] the theorem is stated for projective curves. The transition between the cases, presented in subsection B.7, is completely standard. For this purpose, the following definitions enable us to relate an affine variety with its 'corresponding' projective variety. First we need the following definitions.

Definition B.11 (homogenization).

- *For a polynomial $f \in \mathbb{F}[x_1, \ldots, x_n]$ of degree d, we define the homogenized version $f^h \in \mathbb{F}[x_0, \ldots, x_n]$ by*

$$f^h(x_0, x_1, \ldots, x_n) = x_0^d \cdot f(x_1/x_0, \ldots, x_n/x_0).$$

- *Similarly, for an ideal $I = < f_1, \ldots, f_s >$ we define the ideal $I^h = < f^h | f \in I >$. Note that I^h is always a homogenous. In particular, it is easy to see that $I^h =< f_1^h, \ldots, f_s^h >$.*

We can now define the projective closure of an affine variety.

Definition B.12 (projective closure). *Let $V \subseteq \mathbb{F}^n$ be an affine variety with ideal $I = \mathbf{I}(V)$. We define the* projective closure *$\overline{V} \subseteq \mathbb{P}^n$ to be the projective variety $\mathbf{V}(I^h)$. Let $U_0 \subseteq \mathbb{P}^n$ be defined as $U_0 = \{(a_0, a_1, \ldots, a_n) \in \mathbb{P}^n | a_0 = 1\}$. Note that U_0 can be identified with $\mathbb{F}^n$. Thus, we can think of an affine variety $V \subseteq \mathbb{F}^n$ as being contained in U_0. For a projective variety $V \subseteq \mathbb{P}^n$, we denote $V^a \triangleq V \cap U_0$. Intuitively, this is "the affine part of V".*

The following propositions show various connections between an affine variety and its projective closure.

Proposition B.4 ([17] Chapter 8, §4, Proposition 7 and Exercise 9). *Let $V \subseteq \mathbb{F}^n$ be an affine variety. Then*

1. *$\overline{V} \cap U_0 = V$.*
2. *V is irreducible if and only if $\overline{V}$ is irreducible.*

Proposition B.5 ([17] Chapter 9, §3, Theorem 12). *Let $V \subseteq \mathbb{F}^n$ be an affine variety. Then*

$$dim(V) = dim(\overline{V}).$$

Proposition B.6 ([17] Chapter 8, §4, Theorem 8). *Let $f_1, \ldots, f_r \in \mathbb{F}[x_1, \ldots, x_n]$ be polynomials such that $V = \mathbf{V}(f_1, \ldots, f_r) \subseteq \mathbb{F}^n$ is non-empty. Then*

$$\overline{V} = \mathbf{V}(f_1^h, \ldots, f_r^h).$$

Claim B.12.1. *Let $V_1, \ldots, V_r \subseteq \mathbb{F}^n$ be affine varieties. Then $\overline{V_1 \cup \ldots \cup V_r} = \overline{V}_1 \cup \ldots \cup \overline{V}_r$.*

Proof. We prove the claim for $r = 2$. The statement for general r follows by induction.

Let I_1, I_2 be the ideals $\mathbf{I}(V_1), \mathbf{I}(V_2)$ respectively. It can be shown that $\mathbf{V}(I_1^h \cap I_2^h) = \mathbf{V}((I_1 \cap I_2)^h)$. We have

$$\overline{V_1 \cup V_2} = \mathbf{V}((I_1 \cap I_2)^h) = \mathbf{V}(I_1^h \cap I_2^h) = \mathbf{V}(I_1^h) \cup \mathbf{V}(I_2^h) = \overline{V}_1 \cup \overline{V}_2,$$

where we used Proposition B.2 in the first and second to last equalities. □

Corollary B.1. *Let $V \subseteq \mathbb{F}^n$ be an affine variety with irreducible components $V_1, \ldots, V_r$. Then, the irreducible components of $\overline{V} \subseteq \mathbb{P}^n$ are $\overline{V}_1, \ldots, \overline{V}_r$.*

Proof. Follows from Proposition B.4 and Claim B.12.1. □

Claim B.12.2. *Let $V \subseteq \mathbb{F}^n$ be an affine variety. If $f \in \mathbb{F}[x_1, \ldots, x_n]$ does not vanish identically on V then f^h does not vanish identically on $\overline{V} \subseteq \mathbb{P}^n$.*

Proof. For any $a \in \mathbb{F}^n$, $f(a) = f^h(1, a)$. Therefore, if $f(a) \neq 0$ for $a \in V$, then $f^h(1, a) \neq 0$, where $(1, a) \in \overline{V}$ by Proposition B.4. □

B.5 The Dimension of Intersections of Hypersurfaces

We say that an affine (projective) variety V is a *hypersurface* if $V = \mathbf{V}(f)$ for a (homogenous) polynomial f. In this subsection we state and prove standard results regarding the dimension of intersections of hypersurfaces. The following definition will be important.

Definition B.13. *We say that an affine or projective variety V has* pure dimension *if all its irreducible components have the same dimension.*

We need the following propositions about the intersection of a hypersurface with a variety.

Proposition B.7 ([17] Chapter 9, §4, Proposition 7)**.** *Let $V \subseteq \mathbb{P}^n$ be a projective variety with $dim(V) \geq 1$. Then for any non-constant homogenous polynomial $f \in \mathbb{F}[x_0, \ldots, x_n]$, $V \cap \mathbf{V}(f) \neq \emptyset$.*

Proposition B.8 ([61], Chapter I, §6, Corollary 1 of Theorem 5)**.** *Let $V \subseteq \mathbb{P}^n$ be an irreducible projective variety. Let $f \in \mathbb{F}[x_0, \ldots, x_n]$ be a homogenous polynomial that does not vanish identically on V and denote $H = \mathbf{V}(f)$. If $V \cap H \neq \emptyset$, then $V \cap H$ has pure dimension $dim(V) - 1$.*

Claim B.13.1. *Let $V \subseteq \mathbb{P}^n$ be a projective variety of pure dimension $dim(V) \geq 1$. Let $f \in \mathbb{F}[x_0, \ldots, x_n]$ be a non-constant homogenous polynomial and let $H = \mathbf{V}(f) \subseteq \mathbb{P}^n$. Assume that f does not vanish identically on any of the irreducible components of V. Then $V \cap H$ has pure dimension $dim(V) - 1$.*

Proof. Let $V = Z_1 \cup \ldots \cup Z_k$ be the decomposition of V into irreducible components. Fix any $j \in [k]$. By Proposition B.7, $Z_j \cap H$ is non-empty, and since f does not vanish on Z_j, by Proposition B.8 all irreducible components of $Z_j \cap H$ have dimension $dim(V) - 1$. To conclude, note that the union of the irreducible components of $Z_j \cap H$ over all $j \in [k]$ is $V \cap H$, and therefore the irreducible components of $V \cap H$ are just a subset of these components (excluding any component that is contained in another). Hence, all irreducible components of $V \cap H$ have dimension $dim(V) - 1$ and the claim follows. □

As a special case we get the following.

Corollary B.2. *Let $f \in \mathbb{F}[x_0, \ldots, x_n]$ be a non-constant homogenous polynomial. Then the hypersurface $H = \mathbf{V}(f) \subseteq \mathbb{P}^n$ has pure dimension $n - 1$.*

Proof. $\mathbb{P}^n$ can be shown to be irreducible and in particular has pure dimension. Thus, using Claim B.13.1 with $V = \mathbb{P}^n$ we get the desired result. □

We can now state and prove the main lemma we use regarding the dimension of intersections of hypersurfaces.

Lemma B.9. *Let $0 < r < n$ be integers and let $f_1, \ldots, f_r \in \mathbb{F}[x_0, \ldots, x_n]$ be non-constant homogenous polynomials. For each $i \in [r]$, let $H_i = \mathbf{V}(f_i) \subseteq \mathbb{P}^n$ and $V_i = \mathbf{V}(f_1, \ldots, f_i) = H_1 \cap \ldots \cap H_i$. Then*

1. *All irreducible components of the projective variety V_r have dimension at least $n - r$.*
2. *Suppose furthermore that for each $2 \leq i \leq r$, f_i does not vanish identically on any of the irreducible components of V_{i-1}. Then V_r is a projective variety of pure dimension $n - r$.*

Proof. We prove the first item by induction on r. For $r = 1$ this follows from Corollary B.2. Assume the claim for $r - 1$. Let $V_{r-1} = Z_1 \cup \ldots \cup Z_k$ be the decomposition of V_{r-1} into irreducible components. Fix any $j \in [k]$. Similarly to the proof of Claim B.13.1, we will show that all the irreducible components of $Z_j \cap H_r$ have dimension at least $n - r$, and since the irreducible components of V_r are a subset of these, the claim follows. From the induction hypothesis we have $dim(Z_j) \geq n - (r - 1)$. If f_r vanishes on Z_j then $Z_j \cap H_r = Z_j$ (which is the only irreducible component) and we are done. Otherwise, by Claim B.13.1 all components of $Z_j \cap H_r$ have dimension at least $n - r$.
We now prove the second item by induction on r. For $r = 1$ this is exactly Corollary B.2. Assume the claim for $r - 1$. Then by the induction hypothesis, V_{r-1} has pure dimension $n - r + 1$. Therefore, by Claim B.13.1 $V_r = V_{r-1} \cap H_r$ has pure dimension $n - r$. □

We also need the corresponding lemma in affine space.

Lemma B.10. *Let $0 < r < n$ be integers and let $f_1, \ldots, f_r \in \mathbb{F}[x_1, \ldots, x_n]$ be non-constant polynomials. For each $i \in [r]$, let $H_i = \mathbf{V}(f_i) \subseteq \mathbb{F}^n$ and let $V_i = \mathbf{V}(f_1, \ldots, f_i) = H_1 \cap \ldots \cap H_i$. Suppose that for each $2 \leq i \leq r$, f_i does not vanish identically on any of the irreducible components of the affine variety V_{i-1}. Then, if V_r is non-empty it is an affine variety of pure dimension $n - r$.*

Proof. For $1 \leq i \leq r$, let $X_i = \mathbf{V}(f_1^h, \ldots, f_i^h)$. By Proposition B.6, for every $1 \leq i \leq r$ $X_i = \overline{V}_i$. Therefore, by Corollary B.1 the irreducible components of X_{i-1} are simply the projective closures of the irreducible components of V_{i-1}. By Claim B.12.2 it follows that f_i^h does not vanish identically on any of the irreducible components of X_{i-1}. Hence, we can use Lemma B.9, and X_r is a projective variety of pure dimension $n - r$; and since $X_r = \overline{V}_r$, using Proposition B.5 V_r is an affine variety of pure dimension $n - r$. □

B.6 The Degree of Intersections of Hypersurfaces

We now discuss degree. The main result we prove is the following corollary of Bezout's theorem.

Lemma B.11. *Let $f_1, \ldots, f_r \in \mathbb{F}[x_0, \ldots, x_n]$ be non-constant homogenous polynomials of degrees $d_1, \ldots, d_r$ respectively, and let $D = d_1 \cdots d_r$. Let $X = \mathbf{V}(f_1, \ldots, f_r) \subseteq \mathbb{P}^n$. Assume that $dim(X) = n - r$. Then*

1. *$deg(X) \leq D$.*
2. *The number of irreducible components of X is at most D.*

Using this Lemma, we immediately get a bound on the number of irreducible components of an affine variety.

Lemma B.12. *Let $f_1, \ldots, f_r \in \mathbb{F}[x_1, \ldots, x_n]$ be non-constant polynomials of degrees $d_1, \ldots, d_r$, respectively, and let $D = d_1 \cdots d_r$. Let $V = \mathbf{V}(f_1, \ldots, f_r) \subseteq \mathbb{F}^n$. Assume that V is non-empty and $dim(V) = n - r$. Then the number of irreducible components of V is at most D.*

Proof. Let $X = \overline{V}$. By Proposition B.6, $X = \mathbf{V}(f_1^h, \ldots, f_r^h)$. Therefore, by Lemma B.11, X has at most D irreducible components, and by Corollary B.1 V has at most D irreducible components. □

The following proposition states that a degree of a hypersurface is at most the degree of any polynomial defining it.

Proposition B.13 ([17], Chapter 9, §4, Exercise 12). *Let f be a non-constant homogenous polynomial. Let $H = \mathbf{V}(f_1)$. Then $deg(H) \leq deg(f)$.*

We will need the following definitions taken from [36].

Definition B.14. *Let $X, Y \subseteq \mathbb{P}^n$ be projective varieties. We say that X and Y* intersect properly *if*

$$dim(X \cap Y) = dim(X) + dim(Y) - n.$$

We quote (a corollary of) Bezout's theorem.

Theorem B.2 (Bezout-[36] Chapter 18, Theorem 18.4 and Corollary 18.5). *Let $X, Y \subseteq \mathbb{P}^n$ be projective varieties of pure dimension intersecting properly. Then*

1. *$deg(X \cap Y) \leq deg(X) \cdot deg(Y)$.*
2. *The number of irreducible components of $X \cap Y$ is at most $deg(X) \cdot deg(Y)$.*

Claim B.14.1. *Let $X = \mathbf{V}(f_1, \ldots, f_r) \subseteq \mathbb{P}^n$ where $f_1, \ldots, f_r \in \mathbb{F}[x_0, \ldots, x_n]$ are non-constant homogenous polynomials. Assume that $dim(X) = n - r$. For $i = 1, \ldots, r$ let $H_i = \mathbf{V}(f_i)$ and $X_i = \mathbf{V}(f_1, \ldots, f_i) = H_1 \cap \ldots \cap H_i$. Then for all $i \in [r]$, X_i has pure dimension $n - i$.*

Proof. By Lemma B.9, all irreducible components of $\mathbf{V}(f_1, \ldots, f_i)$ have dimension at least $n - i$. Thus, it is enough to prove that $\mathbf{V}(f_1, \ldots, f_i)$ has (not necessarily pure) dimension $n - i$. We prove this by backwards induction on i. For $i = r$ it is already given that $dim(X) = dim(X_r) = n - r$. Assume the claim for $i + 1$ holds and assume for contradiction that $dim(X_i) \neq n - i$. Using Lemma B.9 it follows that $dim(X_i) > n - i$. Therefore, by Claim B.13.1 $dim(X_{i+1}) = dim(X_i \cap \mathbf{V}(f_{i+1})) > n - (i + 1)$, and this contradicts the induction hypothesis. □

We can now prove Lemma B.11.

Proof. (of Lemma B.11). We prove the claim by induction on r. For $r = 1$, it follows from Proposition B.13 that $deg(X) \leq deg(f_1) = d_1$. Assume the claim for $r - 1$. For $i = 1, \ldots, r$ denote $H_i = \mathbf{V}(f_i)$. Given $H_1, \ldots, H_r$, let $X_{r-1} = H_1 \cap \ldots \cap H_{r-1}$. We know from the induction hypothesis that

$$deg(X_{r-1}) \leq d_1 \cdots d_{r-1}.$$

From Claim B.14.1, X_{r-1} has pure dimension $n - (r - 1)$ and it follows that X_{r-1} and H_r intersect properly. Therefore, we can use Theorem B.2 and get

$$\deg(X) = deg(X_{r-1} \cap H_r) \leq deg(X_{r-1}) \cdot \deg(H_r) \leq d_1 \cdots d_r = D.$$

Similarly, from Theorem B.2 we get that the number of irreducible components of X is at most $deg(X_{r-1}) \cdot deg(H_r) \leq D$. □

B.7 Bombieri's Theorem

We quote an estimate of Bombieri [8] for character sums over projective curves and show that the estimate can be used also for affine curves. (Recall that a curve is a variety of dimension 1.) First we introduce some notation. Let $X \subseteq \mathbb{P}^n$ be a projective curve of degree D. Let $\mathbb{F}$ denote the algebraic closure of $\mathbb{F}_p$ for some prime p. Let $R \in \mathbb{F}_p(x_0, \ldots, x_n)$ be a homogenous rational function whose numerator and denominator both have degree d. Then, for any $x \in \mathbb{F}^{n+1}$ and non-zero $\lambda \in \mathbb{F}$ we have

$$R(\lambda \cdot x) = \frac{p(\lambda \cdot x)}{q(\lambda \cdot x)} = \frac{\lambda^d p(x)}{\lambda^d q(x)}$$

$$= \frac{p(x)}{q(x)} = R(x).$$

Therefore R is a well-defined function on points of $\mathbb{P}^n$ that are not poles of R, i.e., points $x \in \mathbb{P}^n$ such that $q(x) \neq 0$. We define

$$S_m(R, X) \triangleq \sum_{x \in X_m, q(x) \neq 0} e_p(\sigma R(x))$$

where X_m is the set of points of X with coordinates in $\mathbb{F}_{p^m}$, σ denotes the trace[4] from $\mathbb{F}_{p^m}$ to $\mathbb{F}_p$ and $e_p(x)$ is the function $e^{2\pi i x/p}$. Note that we sum only over non-poles of R.

Theorem B.3 (Theorem 6 in [8]). *Let R and X be as above. Let $\Gamma_1, \ldots, \Gamma_L$ be the irreducible components of X. Assume R is non-constant on Γ_i for $i = 1, \ldots, L$. If $d \cdot D < p$ then*

$$|S_m(R, X)| \leq 4dD^2 \cdot p^{m/2}.$$

For an affine curve $C \subseteq \mathbb{F}^n$ and a polynomial $g \in \mathbb{F}_p[x_1, \ldots, x_n]$ we define

$$S_m(g, C) \triangleq \sum_{(a_1, \ldots, a_m) \in C_m} e_p(\sigma g(a_1, \ldots, a_m))$$

where C_m denotes the set of points of C with coordinates in $\mathbb{F}_{p^m}$. We also denote $S(g, C) \triangleq S_1(g, C)$. We can now state and prove a version of Theorem B.3 for affine curves.

Theorem B.4. *Let $V \subseteq \mathbb{F}^n$ be an affine curve such that $V = \mathbf{V}(f_1, \ldots, f_{n-1})$ for polynomials $f_i \in \mathbb{F}[x_1, \ldots, x_n]$. Let $D = deg(f_1) \cdots deg(f_{n-1})$. Let $V_1, \ldots, V_L$ be the irreducible components of V. Let $g \in \mathbb{F}_p[x_1, \ldots, x_n]$ be a polynomial of degree d that is non-constant on some V_i. Let C be the union*

[4] See [39] for a definition of the trace function. For the case $m = 1$, which is the only one we will use, the trace is simply the identity function.

of the irreducible components V_i such that g is non-constant on V_i. Assume that $d \cdot D < p$. We have

$$S_m(g, C) \leq 4dD^2 \cdot p^{m/2}.$$

In particular,

$$S(g, C) \leq 4dD^2 \cdot p^{1/2}.$$

Proof. We identify g with a homogenous rational function R defined as

$$R(x_0, x) = \frac{g^h(x_0, x)}{x_0^d}.$$

Note that for every $a \in \mathbb{F}^n$ $R(1, a) = g(a)$.
Denote $X = \overline{C}$.

Claim B.14.2.

$$S_m(g, C) = S_m(R, X).$$

Proof. Using Proposition B.4 X consists precisely of the points $(1, a)$ where $a \in C$ and, possibly, some 'points at infinity', i.e., points of the form $(0, a)$ for $a \in \mathbb{F}^n$. Since R has poles on all points of the form $(0, a)$ and $R(1, a) = g(a)$ for all $x \in \mathbb{F}^n$, we get that summing R over all non-poles in X is exactly the same as summing g over all of C. In particular, summing R over all non-poles in X_m is exactly the same as summing g over all of C_m. That is,

$$S_m(g, C) = S_m(R, X).$$

□

We now want to bound $S_m(R, X)$ using Theorem B.3. Note that both the numerator and the denominator of R are homogenous of degree exactly d, so R is suitable for the theorem. We need to show that X is a projective variety of dimension 1 such that R is non-constant on any of its irreducible components: Recall that the irreducible components of C are simply a subset of $V_1, \ldots, V_L$. Assume w.l.o.g. that $C = V_1 \cup \ldots \cup V_r$. Using Corollary B.1, it is clear that if g is non-constant on the irreducible components $V_1, \ldots, V_r$ of C, then R is non-constant on the irreducible components $\overline{V}_1, \ldots, \overline{V}_r$ of X. By Proposition B.5 and Corollary B.1 $dim(\overline{V}) = 1$ and $\overline{V}_1, \ldots, \overline{V}_L$ are the irreducible components of $\overline{V}$. By Proposition B.6, $\overline{V} = \mathbf{V}(f_1^h, \ldots, f_{n-1}^h)$, and therefore by Claim B.14.1 for every i $\overline{V}_i$ has dimension 1. It follows that $X = \overline{V}_1 \cup \ldots \cup \overline{V}_r$ has dimension 1.

Finally, we need to bound the degree of X. By Lemma B.11 $deg(\overline{V}) \leq D$. Since the degree of a projective variety is the sum of degrees of its irreducible components (see [36], Chapter 18), $deg(X) \leq D$.

Therefore, we can use Theorem B.3. We get

$$|S_m(g, C)| = |S_m(R, X)| \leq 4dD^2 \cdot p^{m/2}.$$

□

Bibliography

[1] N. Alon. Tools from higher algebra. In *R. L. Graham & M. Grotschel & L. Lovasz (eds.), Handbook of Combinatorics, Elsevier and The MIT Press*, volume 2. 1995.

[2] N. Alon, O. Goldreich, J. Håstad, and R. Peralta. Simple constructions of almost k-wise independent random variables. In *Proceedings of the 31st Annual IEEE Symposium on Foundations of Computer Science*, volume II, pages 544–553, 1990.

[3] B. Barak, R. Impagliazzo, and A. Wigderson. Extracting randomness using few independent sources. *SIAM J. Comput*, 36(4):1095–1118, 2006.

[4] B. Barak, G. Kindler, R. Shaltiel, B. Sudakov, and A. Wigderson. Simulating independence: New constructions of condensers, Ramsay graphs, dispersers, and extractors. In *Proceedings of the 37th Annual ACM Symposium on Theory of Computing*, pages 1–10, 2005.

[5] B. Barak, A. Rao, R. Shaltiel, and A. Wigderson. 2-source dispersers for sub-polynomial entropy and Ramsey graphs beating the Frankl–Wilson construction. In *Proceedings of the 38th Annual ACM Symposium on Theory of Computing*, pages 671–680, 2006.

[6] M. Bellare and J. Rompel. Randomness-efficient oblivious sampling. FOCS 1994.

[7] M. Ben-Or and N. Linial. Collective coin flipping. *ADVCR: Advances in Computing Research*, 5:91–115, 1989.

[8] E. Bombieri. On exponential sums in finite fields. *American Journal of Mathematics*, 88:71–105, 1966.

[9] J. Bourgain. More on the sum-product phenomenon in prime fields and its applications. *International Journal of Number Theory*, 1:1–32, 2005.

[10] J. Bourgain. On the construction of affine extractors. *Geometric And Functional Analysis*, 17(1):33–57, 2007.

[11] V. Boyko. On the security properties of OAEP as an all-or-nothing transform. In *Proc. 19th International Advances in Cryptology Conference – CRYPTO '99*, pages 503–518, 1999.

[12] R. Canetti, Y. Dodis, S. Halevi, E. Kushilevitz, and A. Sahai. Exposure-resilient functions and all-or-nothing transforms. *Lecture Notes in Computer Science*, 1807, 2000.

[13] I. L. Carter and M. N. Wegman. Universal classes of hash functions. In *Proceedings of the 9th Annual ACM Symposium on Theory of Computing*, pages 106–112, 1977.

[14] B. Chor and O. Goldreich. Unbiased bits from sources of weak randomness and probabilistic communication complexity. *SIAM Journal on Computing*, 17(2):230–261, April 1988. Special issue on cryptography.

[15] B. Chor, O. Goldreich, J. Hastad, J. Friedman, S. Rudich, and R. Smolensky. The bit extraction problem or t-resilient functions. In *Proceedings of the 26th Annual IEEE Symposium on Foundations of Computer Science*, pages 396–407, 1985.

[16] A. Cohen and A. Wigderson. Dispersers, deterministic amplification and weak random sources. In *Proceedings of the 30th Annual IEEE Symposium on Foundations of Computer Science*, pages 14–25, 1989.

[17] D. Cox, J. Little, and D. O'Shea. *Ideals, Varieties and Algorithms.* Springer, 1992.

[18] Y. Dodis. *Exposure-Resilient Cryptography.* PhD thesis, Department of Electrical Engineering and Computer Science, MIT, August 2000.

[19] Y. Dodis, A. Elbaz, R. Oliveira, and R. Raz. Improved randomness extraction from two independent sources. In *RANDOM: International Workshop on Randomization and Approximation Techniques in Computer Science*, pages 334–344, 2004.

[20] Y. Dodis, A. Sahai, and A. Smith. On perfect and adaptive security in exposure-resilient cryptography. *Lecture Notes in Computer Science*, 2045, 2001.

[21] R. Ehrenborg and G. Rota. Apolarity and canonical forms for homogeneous polynomials. *Europ. J. Combinatorics*, 14:157–181, 1993.

[22] A. Elbaz. Improved constructions for extracting quasi-random bits from sources of weak randomness. *M.Sc. Thesis, Weizmann Institute*, 2003.

[23] S. Even, O. Goldreich, M. Luby, N. Nisan, and B. Velickovic. Efficient approximation of product distributions. *RSA: Random Structures & Algorithms*, 13, 1998.

[24] A. Fiat and M. Naor. Implicit O(1) probe search. *SICOMP: SIAM Journal on Computing*, 22, 1993.

[25] A. Gabizon and R. Raz. Deterministic extractors for affine sources over large fields. In *Proceedings of the 46th Annual IEEE Symposium on Foundations of Computer Science*, pages 407–418, 2005.

[26] A. Gabizon, R. Raz, and R. Shaltiel. Deterministic extractors for bit-fixing sources by obtaining an independent seed. *SICOMP: SIAM Journal on Computing*, 36(4):1072–1094, 2006.

[27] A. Gabizon and R. Shaltiel. Increasing the output length of zero-error dispersers. In *APPROX-RANDOM*, pages 430–443, 2008.

[28] F. R. Gantmacher. *The Theory of Matrices*, volume 1. New York, NY, 1959.

[29] O. Goldreich. Three XOR-lemmas - an exposition. *Electronic Colloquium on Computational Complexity (ECCC)*, 2(056), 1995.

[30] O. Goldreich. A sample of samplers – A computational perspective on sampling (survey). In *ECCCTR: Electronic Colloquium on Computational Complexity, technical reports*, 1997a.

[31] O. Goldreich. A sample of samplers – A computational perspective on sampling (survey). In *ECCCTR: Electronic Colloquium on Computational Complexity, Technical Reports*, 1997b.

[32] R. L. Graham, B. L. Rothschild, and J. H. Spencer. *Ramsey Theory.* Wiley, 1980.

[33] R. L. Graham and J. H. Spencer. A constructive solution to a tournament problem. *Canad. Math. Bull.*, 14:45–48, 1971.

[34] V. Guruswami, C. Umans, and S. P. Vadhan. Unbalanced expanders and randomness extractors from Parvaresh-Vardy codes. In *Proceedings of the 22nd Annual IEEE Conference on Computational Complexity*, pages 96–108, 2007.

[35] A. Hales and R. Jewett. Regularity and positional games. *Trans. Amer. Math. Soc.*, 106:222–229, 1963.

[36] J. Harris. *Algebraic Geometry - A First Course.* Springer, 1992.

[37] J. Kamp and D. Zuckerman. Deterministic extractors for bit-fixing sources and exposure-resilient cryptography. *SIAM J. Comput*, 36(5):1231–1247, 2007.

[38] N. Kayal. The complexity of the annihilating polynomial. *Manuscript*, 2007.

[39] R. Lide and H. Niederreiter. *Finite fields.* Cambridge University Press, New York, NY, USA, 1997.

[40] R. Lipton and N. Vishnoi. Manuscript. 2004.

[41] L. Lovasz. *Combinatorial Problems and Exercises.* North-Holland, Amsterdam, 1979.

[42] C. Lu, O. Reingold, S. Vadhan, and A. Wigderson. Extractors: Optimal up to constant factors. In *Proceedings of the 35th Annual ACM Symposium on Theory of Computing*, pages 602–611, 2003.

[43] M. S. L'vov. Calculation of invariants of programs interpreted over an integrality domain. *Kibernetika*, (4):23–28, 1984.

[44] J. Naor and M. Naor. Small-bias probability spaces: Efficient constructions and applications. In *Proceedings of the 22nd Annual ACM Symposium on Theory of Computing*, pages 213–223, 1990.

[45] M. Naor, A. Nussboim, and E. Tromer. Efficiently constructible huge graphs that preserve first order properties of random graphs. In *TCC*, pages 66–85, 2005.

[46] N. Nisan and A. Ta-Shma. Extracting randomness: A survey and new constructions. *Journal of Computer and System Sciences*, 58, 1999.

[47] N. Nisan and D. Zuckerman. More deterministic simulation in logspace. In *Proceedings of the Twenty-Fifth Annual ACM Symposium on Theory of Computing*, pages 235–244, New York, NY, USA, 1993. ACM Press.

[48] N. Nisan and D. Zuckerman. Randomness is linear in space. *Journal of Computer and System Sciences*, 52(1):43–52, 1996.

[49] J. Radhakrishnan and A. Ta-Shma. Bounds for dispersers, extractors, and depth-two superconcentrators. *SIAM Journal on Discrete Mathematics*, 13(1):2–24, 2000.

[50] A. Rao. Extractors for a constant number of polynomially small min-entropy independent sources. In *Proceedings of the 38th Annual ACM Symposium on Theory of Computing*, pages 497–506, 2006.

[51] A. Rao. An exposition of Bourgain's 2-source extractor. Technical Report TR07-034, ECCC, 2007.

[52] A. Rao. Extractors for low weight affine sources. *Manuscript*, 2008.

[53] R. Raz. Extractors with weak random seeds. In *Proceedings of the 37th Annual ACM Symposium on Theory of Computing*, pages 11–20, 2005.

[54] R. Raz, O. Reingold, and S. Vadhan. Error reduction for extractors. In *Proceedings of the 40th Annual IEEE Symposium on Foundations of Computer Science*, 1999.

[55] R. Raz, O. Reingold, and S. Vadhan. Extracting all the randomness and reducing the error in Trevisan's extractors. In *Proceedings of the 31st Annual ACM Symposium on Theory of Computing*, pages 149–158, 1999.

[56] O. Reingold, R. Shaltiel, and A. Wigderson. Extracting randomness via repeated condensing. In *Proceedings of the 41st Annual IEEE Symposium on Foundations of Computer Science*, 2000.

[57] R. Rivest. All-or-nothing encryption and the package transform. In *Fast Software Encryption: 4th International Workshop, FSE*, volume 1267 of *Lecture Notes in Computer Science*, 1997.

[58] M. Santha and U. V. Vazirani. Generating quasi-random sequences from semi-random sources. *Journal of Computer and System Sciences*, 33:75–87, 1986.

[59] W. M. Schmidt. *Equations over Finite Fields: An Elementary Approach*, volume 536. Springer-Verlag, Lecture Notes in Mathematics, 1976.

[60] J. T. Schwartz. Fast probabilistic algorithms for verification of polynomial identities. *J. ACM*, 27(4):701–717, 1980.

[61] I. R. Shafarevich. *Basic algebraic geometry.* Springer, 1994.

[62] R. Shaltiel. Recent developments in explicit constructions of extractors. *Bulletin of the EATCS*, 77:67–95, 2002.

[63] R. Shaltiel. How to get more mileage from randomness extractors. In *CCC '06: Proceedings of the 21st Annual IEEE Conference on Computational Complexity*, pages 46–60, 2006.

[64] R. Shaltiel and C. Umans. Simple extractors for all min-entropies and a new pseudo-random generator. In *Proceedings of the 42nd Annual IEEE Symposium on Foundations of Computer Science*, 2001.

[65] A. Ta-Shma, D. Zuckerman, and S. Safra. Extractors from Reed-Muller codes. In IEEE, editor, *Proceedings of the 42nd Annual IEEE Symposium on Foundations of Computer Science*, pages 638–647, 2001. IEEE Computer Society Press.

[66] L. Trevisan. Construction of extractors using pseudorandom generators. In *Proceedings of the 31st ACM Symposium on Theory of Computing*, 1999.

[67] L. Trevisan and S. Vadhan. Extracting randomness from samplable distributions. In *Proceedings of the 41st Annual IEEE Symposium on Foundations of Computer Science*, 2000a.

[68] L. Trevisan and S. Vadhan. Extracting randomness from samplable distributions. In *Proceedings of the 41st Annual Symposium on Foundations of Computer Science*, pages 32–42, 2000b.

[69] S. Vadhan. On constructing locally computable extractors and cryptosystems in the bounded storage model, November 01 2002.

[70] U. Vazirani. Strong communication complexity or generating quasi-random sequences from two communicating semi-random sources. *Combinatorica*, 7:375–392, 1987.

[71] J. von Neumann. Various techniques used in connection with random digits. *Applied Math Series*, 12:36–38, 1951.

[72] A. Weil. On some exponential sums. In *Proc. Nat. Acad. Sci. USA*, volume 34, pages 204–207, 1948.

[73] T. Wooley. A note on simultaneous congruences. *J. Number Theory*, 58:288–297, 1996.

[74] M. Prabhakaran Y. Dodis, S. J. Ong and A. Sahai. On the (im)possibility of cryptography with imperfect randomness. In *FOCS 2004*, 2004.

[75] A. C.-C. Yao. Should tables be sorted? *J. ACM*, 28(3):615–628, 1981.

[76] R. Zippel. Probabilistic algorithms for sparse polynomials. In *Proceedings of the International Symposium on Symbolic and Algebraic Computation*, pages 216–226. Springer-Verlag, 1979.